EL NIÑO 1997–1998

EL NIÑO
1997–1998

The Climate Event of the Century

Edited by
STANLEY A. CHANGNON

Contributors
Gerald D. Bell, David Changnon, Stanley A. Changnon,
Vernon E. Kousky, Roger A. Pielke Jr., and Lee Wilkins

OXFORD
UNIVERSITY PRESS

2000

OXFORD
UNIVERSITY PRESS

Oxford   New York
Athens   Auckland   Bangkok   Bogotá   Buenos Aires   Calcutta
Cape Town   Chennai   Dar es Salaam   Delhi   Florence   Hong Kong   Istanbul
Karachi   Kuala Lumpur   Madrid   Melbourne   Mexico City   Mumbai
Nairobi   Paris   São Paulo   Singapore   Taipei   Tokyo   Toronto   Warsaw

and associated companies in
Berlin   Ibadan

Library of Congress Cataloging-in-Publication Data
El Niño 1997–1998 : the climate event of the century / edited by
Stanley A. Changnon : contributors, Gerald D. Bell . . . [et al.].
p.   cm.
Includes bibliographical references and index.
ISBN 0-19-513551-2; ISBN 0-19-513552-0 (pbk.)
1. El Niño Current.   2. United States—Climate.
3. Climatic changes—United States.   I. Changnon, Stanley Alcide.
II. Bell, Gerald D.
GC296.8.E4E54   2000
551.6—dc21   99-35619

9 8 7 6 5 4 3 2

Printed in the United States of America
on acid-free paper

This book is dedicated to those scientists who have developed today's advanced understanding of how the tropical Pacific influences the atmosphere and changes the climate conditions over North America.

PREFACE

This book represents a conjunction of the talents and interests needed to obtain a reasonably in-depth perspective of a nationally high visibility and complex climate event like El Niño 97–98. The book's focus is on what happened in the United States where El Niño became a household word and scientifically the "climate event of the century." The book is a "snapshot" of events during a 14-month period starting when El Niño developed in the late spring of 1997 and ending by early summer 1998. Some material was issued after June 1998, but it is tied to activities during the event.

The book has a strong focus on information that appeared in the media and on the Internet—these were the two major sources of information, good or bad, correct or not, issued during the event. What most everyone learned came from newspapers, television, and the Internet collectively comprising a myriad of sources.

The concept of such a study was the brainchild of Lee Wilkins and the book's editor, Stan Changnon. Both share interests and concerns over how the mass media handle and interpret scientific information and they had worked together on a book about the great Midwestern flood of 1993. The scientific anomaly that El Niño 97–98 represented and its massive press coverage led them to conceive of a project that would describe and try to interpret the entirety of El Niño in the U.S. This concept was shared with Jim Laver, the Assistant Director of the Climate Prediction Center (CPC), and this led to a project funded by CPC to perform the analysis. Lee assessed the media's handling of the issue, and Stan addressed the scientific issues that arose and the societal impacts that were created from the El Niño weather.

Other talents were needed to accomplish the goals of the project and the book envisioned. One involved extensive sampling of users of the long-range weather predictions issued based on El Niño's teleconnections to U.S. weather patterns. David Changnon, who had performed prediction-user studies before, agreed to study this issue and another piece of the project puzzle was in place.

Another key area of interest concerned analysis of the policy responses and ramifications of El Niño and its predictions. The team was fortunate to get Roger A. Pielke Jr., a political scientist with a deep interest in weather issues, to handle this phase of the project.

The final need was for atmospheric scientists with hands-on expertise regarding El Niño events to develop a physical description of the phenomenon and focus on what happened in 1997–1998 with regards to the tropical Pacific, weather patterns over the U.S., and the resulting weather predictions. Two such skilled scientists, Vernon Kousky and Gerald Bell of the Climate Prediction Center, agreed to become the final essential part of the team needed to describe the event and its impacts and implications on society.

The creation of a book typically focuses on the potential audience and in fact, most publishers insist on such a focus. The authors' of this book envisioned a diverse audience. One such audience includes university staff and students with interests in natural hazards, atmospheric sciences, science policy, economics, and journalism. Another audience of this book includes decision makers and leaders in weather-sensitive businesses and government agencies. The book contains several recommendations (and lessons learned) that should have particular relevance to certain government agencies. Last but not least is the American public which became enamored with El Niño 97–98. Hopefully, this book contains material that will be of interest and value to all these parties.

Mahomet, Illinois   S. A. C.
March 1999

ACKNOWLEDGMENTS

Any such endeavor requires the assistance and support of many individuals and institutions. We give special thanks to Jim Laver of the Climate Prediction Center (CPC) who enthusiastically supported the concept of this book and helped get the project launched and funded by CPC. We credit Charles Mercer for preparing quality photographic prints. The interest and assistance of several cartoonists including Ed Stein were important contributions. Richard Andrews and Robert Eplett of the California Office of Emergency Services provided timely information and useful photographs. Associates in New England and Alabama contributed useful photographs depicting weather outcomes. Several maps and graphs presented were prepared by the National Climatic Data Center of NOAA, and are found on the Internet, and we are grateful for their use.

The authors of chapter 2 thank Michael Halpert for his help in producing illustrations. Alexander Calhoun, a graduate student at the University of Missouri helped with the content analysis of chapter 3; and the National Association of Science Writers was helpful in allowing Lee Wilkins to survey them. Bob Scott helped analyze the Internet products and their scientific content as presented in chapter 4.

The sampling of users of the forecasts presented in chapter 5 was aided considerably by several individuals who identified potential persons for interviews. Those who helped in this process included Kelly Redmond of the Western Regional Climate Center; Tom Potter, director of the Western Region of National Weather Service; Tom Henderson, president of Atmospherics, Inc., in Fresno, CA; Ken Mielke of the National Weather Service in Monterey, CA; Art De Gaetano of the Northeast Regional Climate Center; Eric Swartz of the South Florida Water Management District; and William Easterling of Pennsylvania State University.

We are also truly indebted to the eighty-seven persons who gave considerable time and thoughts to respond to our interviews, the basis for chapter 5.

The Midwestern Climate Center provided data and considerable information on the Midwestern impacts presented in chapter 6. Members of the Illinois Department of Transportation and Center for Disease Control provided valuable data on lives lost during El Niño, as did Mike Rossetti of the U.S. Department of Transportation. Several others including staff at United Airlines, Schneider Trucking, and Northwest Airlines provided valuable data, and Cynthia P. Cain of First Chicago Leasing provided very useful national economic data.

Conversations with Bill Hooke, Dan Sarewitz, and Rad Byerly were of value in preparing chapter 7; Charles Visserig of FEMA and Craig Fugate of the Florida Department of Emergency Management also provided useful information. The material in chapter 7 reflects a great deal of advice and the teachings of Michael Glantz.

Last but not least, the editor expresses his deep appreciation to Joyce Changnon for her patience and understanding while this document was assembled, and for valuable reviews of the chapters.

CONTENTS

ABBREVIATIONS

AGU   American Geophysical Union
AMS   American Meteorological Society
CDC   Climate Diagnostics Center
CPC   Climate Prediction Center
COAPS   Center for Ocean-Atmosphere Prediction
COLA   Center for Ocean-Land-Atmosphere Studies
DOC   Department of Commerce
ENSO   El Niño Southern Oscillation
FEMA   Federal Emergency Management Agency
IRI   International Research Institute
JPL   Jet Propulsion Laboratory
NASA   National Aeronautics and Space Administration
NOAA   National Oceanic and Atmospheric Administration
OGP   Office of Global Programs
NCAR   National Center for Atmospheric Research
NCEP   National Centers for Environmental prediction
NWS   National Weather Service
PMEL   Pacific Marine Experimental Laboratory
SO   Southern Oscillation
SOI   Southern Oscillation Index
SST   Sea Surface Temperature
USDA   U.S. Department of Agriculture
WMO   World Meteorological Organization

CONTRIBUTORS

Gerald D. Bell is a research meteorologist in the Analysis Branch of NOAA's Climate Prediction Center in Washington, DC. He specializes in monitoring global climate variability, especially that related to El Niño and other large-scale atmospheric processes. He co-authors and co-edits the *Climate Diagnostics Bulletin* and is co-editor of the *El Niño/Southern Oscillation Advisory*. Both are monthly publications that focus on climate issues and El Niño analysis and forecasts, and he is author and editor of *Special Climate Summaries* which provide timely analysis of any major global climate variations. His research interests include El Niño, interannual and intraseasonal climate variations, and large-scale weather patterns in the middle latitudes and tropics.

David Changnon is an atmospheric scientist specializing in climatological studies in the Department of Geography at Northern Illinois University. His major expertise rests in developing climatological information and models for use by weather-sensitive decision makers in agriculture, utilities, insurance, and transportation. He has developed an innovative new "education-to-career" program to train students and simultaneously solve climatological problems facing government and private industry. He has served on various committees of two scientific societies. His research interests also involve the spatial and temporal variability of climate extremes in the U.S.

Stanley A. Changnon is a professor of geography and atmospheric sciences at the University of Illinois at Urbana-Champaign, and chief emeritus of the Illinois State Water Survey. Many of his climatological investigations have addressed extremes like droughts, floods, and severe weather. His interdisciplinary interests include the effects of weather and climate on social and physical systems, government policy, and atmospheric sciences. Five of his papers and a book have won national awards. He is a fellow of three scientific societies and has been president of two professional societies.

Vernon E. Kousky is a research meteorologist in the Analysis Branch of NOAA's Climate Prediction Center in Washington, DC, where he specializes in monitoring global climate variability, especially that related to the El Niño/Southern Oscillation (ENSO). Prior to this position, he was a staff member of the University of Utah and the University of São Paulo in Brazil. He is chief editor of the monthly *Climate Diagnostics Bulletin*. His research interests include interannual and intraseasonal climate variations, the El Niño—Southern Oscillation cycle, and synoptic climatology.

Roger A. Pielke Jr. is a research scientist in the Environmental and Societal Impacts Group at the National Center for Atmospheric Research, and is an Associate Professor of Political Sciences at the University of Colorado. His research expertise relates to the relation of scientific information and public and private sector decision making. His current research areas include societal responses to extreme weather events, policy responses to climate change, and national science policy. He is the co-author of a book on hurricanes and numerous scientific papers.

Lee Wilkins is a professor in the Broadcast News Department at the University of Missouri School of Journalism. Her research focuses on media coverage of environment and risk communication, including studies of media performance and public response to the Bhopal, India, chemical spill, the Chernobyl nuclear disaster, global climate change, and the 1993 Midwest flood. She is a former newspaper reporter and editor who has been teaching journalism and mass communications for 20 years. Her book about the Bhopal incident was selected as a *Choice* outstanding academic book in 1987.

EL NIÑO 1997–1998

1

 ## What Made El Niño 1997–1998 Famous?

The Key Events Associated with a Unique Climatic Event

STANLEY A. CHANGNON

E l Niño 97–98 provided one of the most interesting and widely known climatic events of this century. It garnered enormous attention not only in the scientific community but also in the media and from the American public.

El Niño developed rapidly in the tropical Pacific during May 1997, and by October "El Niño" had become a household phrase across America. Television and radio, newspapers and magazines pummeled America with the dire tales of El Niño during the fall of 1997 as the climate disruption battered the West Coast and the southern United States with storm after storm. Worried families changed vacation plans, and insurance executives pondered losses and raised rates. Victims of every type of severe weather blamed El Niño. After a winter filled with unusual weather, the head of the National Oceanic and Atmospheric Administration (NOAA) declared, "This winter's El Niño ranks as one of the major climatic events of this century." It was the first El Niño observed and forecast from start to finish.

The event was noteworthy from several perspectives.

- First, it became the largest and warmest El Niño to develop in the Pacific Ocean during the past 100 years.
- Second, the news media gave great attention to the event, and El Niño received more attention at all levels than had any previous climate event.
- Third, scientists were able to use El Niño conditions to successfully predict the climate conditions of the winter six months in advance.

- Fourth, the predictive successes brought new credibility to the science of long-range prediction and, in general, acted to increase the public's understanding of the climate and oceanic sciences.
- Fifth, there were notable differences in how weather-sensitive decision makers reacted to the predictions, some used them for great gain, while others, fearing failure, did not.
- Sixth, the great strength of El Niño brought forth claims that the phenomenon was the result of anthropogenic-induced global warming. This possibility was debated and added to the scientific-policy debates surrounding climate change.
- Seventh, the net effect of the El Niño–influenced weather on the United States was an economic benefit, after early fears and predictions of great damages. Some areas, including California and Florida, suffered major losses, but many areas, including the northern United States, realized sizable economic and health-related benefits from the mild winter.

To understand these impacts, it is important to first create and follow a chronology of the major events as they developed during 1997 and 1998. Then we examine the key events and issues, including the societal-economic impacts, the media's attention to the event, the reactions of the public and of policy makers, and the various uses of the El Niño–related climate predictions. Finally the key scientific issue—the debate over the relationship between El Niño and global warming—is analyzed.

A CHRONOLOGY OF MAJOR EVENTS

The following chronology of major events that unfolded during El Niño 97–98 condenses much information, I hope accurately, to allow one to follow the interactions over scientific issues, climate and oceanic predictions, the impacts from the weather, and societal reactions to the event. Throughout the El Niño period, the nation's official prediction agency, the Climate Prediction Center (CPC) of NOAA, issued a series of "ENSO Advisories" (El Niño Southern Oscillation) to report on the latest status of ocean conditions and to issue predictions of what oceanic and climate conditions to expect. Highlights extracted from these advisories are included in the chronology (and appear in italics) with their date of issuance. For example, here is the key information from the two advisories issued just before El Niño developed:

January 1997 Advisory—Sea surface temperatures (SSTs) were predicted to be near normal through mid-1997.

April 9 Advisory—SSTs were up 1°C, but no significance was assigned to this change.

May–July 1997

Instrumented buoys anchored in the tropical Pacific Ocean and satellites of the National Space and Aeronautics Administration began detecting a rapid warming of the ocean waters off the coast of Chile in April and May, a signal that a new El Niño was developing.

> *May 9 Advisory*—Conditions now looked like the beginning stages of a warm episode (El Niño).

By early June, the warming area had grown rapidly, and NOAA scientists began forecasting the impending El Niño. In July its rapid growth, both in areal extent and in depth below the ocean's surface, led oceanographers to predict that this El Niño could become the largest ever.

> *June 10 Advisory*—Further strengthening of the warm episode occurred during May. The SST positive anomalies increased dramatically in the central-east Pacific, becoming the largest temperature anomaly there since August 1983.

> *June 26 Advisory* (note, issued 16 days later)—A strong ENSO had developed, with considerable warming in the first three weeks of June.

In June, NOAA's Climate Prediction Center began issuing long-range predictions for future seasonal climate conditions on the basis of El Niño's influence. Those predictions called for wet and stormy weather on the West Coast and along the southern tier of states and for mild temperatures and below-normal precipitation in the northern two-thirds of the United States during the fall, winter, and early spring. NOAA began to provide an enormous amount of information about El Niño to the news media and on the Internet.

> *July 15 Advisory*—ENSO had continued to strengthen and was now the largest since 1982–1983. Prediction: ENSO conditions to persist through 1997 and into early 1998, the first prediction of duration.

August–September 1997

By August several scientists had declared the 1997 event the largest El Niño in the past 100 years, on the basis of the ocean warming. It closely approximated the magnitude of the 1982–1983 El Niño, which had produced devastating storms and weather losses in California and in the southern states.

> *August 13 Advisory*—Pacific continued to warm. Climate predictions of the effects of ENSO included expected low fall rainfall in Australia and in South America.

A host of scientific research institutions and scientists began issuing their own El Niño–based climate predictions, most in close agreement with the official CPC predictions. Many news releases by federal agencies about El Niño focused on predictions of expected physical impacts. The Federal Emergency Management Agency (FEMA) warned about the dire societal and economic consequences, focusing heavily on the potential for extensive flooding in parts of the United States. Bad smoke pollution in Southeast Asia was blamed on a drought caused by El Niño. Certain scientists, FEMA, NOAA, and some media outlets issued descriptions of the 1982–1983 El Niño and its extensive losses, concluding that the same outcome was in store in 1997–1998. These circumstances led to actions to organize mitigation efforts in California and Florida.

> *September 10 Advisory*—Ocean warming—the most extensive in fifty years—continued through August. Predictions of U.S. winter conditions called for wet weather in the South and warm conditions in the northern states.

Many questions and public fears noted during the early months of El Niño are highlighted in the headlines in Figure 1-1. Most news stories about El Niño cited scientists as sources of the information presented.

October–November 1997

> *October 10 Advisory*—Some ocean areas had the warmest September values in fifty years.

At an El Niño mitigation conference convened in California, Vice President Al Gore stated that the 1997–1998 El Niño was caused by global warming. His assertion fueled a scientific debate that continued throughout El Niño. California and Florida state agencies and some communities launched storm and flood mitigation activities. A few scientists questioned the accuracy of official predictions of the coming winter weather.

The first attribution of a storm to El Niño occurred when a severe and unexpected early winter storm struck the mountainous West and the High Plains late in October. Scientists acknowledged that the absence of Atlantic hurricanes in 1997 was a result of El Niño's influence; then three strong and rather unusual hurricanes developed in the Pacific and struck Mexico late in September and early October. This combination of unusual weather events and associated scientific pronouncements acted to create interest and a belief in El Niño's impact on the nation's weather. These events also provided the first scientific verifications of the accuracy of the fall predictions.

The news media published increasing numbers of predictions of the impacts expected from El Niño–generated weather extremes, offered by a wide variety of scientists, economists, and business persons, as shown in Figure 1-1. The predictions were far from uniform, reflecting a wide lack of understanding of how weather conditions affect complex physical and economic systems. Most of the

What is El Niño ?

The Emergence of El Niño — Potential Implications

California readies itself for El Nino ahead of time

Is another aberrant El Nino

El Niño is coming! It'll drown California! It'll ruin the Olympics! It'll kill seals! It's already making everybody weird!

weather event on the way?

http://www.ogp.noaa.gov/enso/

How Will El Niño Affect the US?		
Climate Forecasts:		Lessons from Past El Niños:
• The next 1 to 3 months		

MARKETING MOVES

Cold Winter On the Way? Some Bet On It

El Nino rules world grain markets

El Nino and its impact on world weather is the biggest factor influencing world crops -- ... today. It's already im-- ... Australia and .

The wrath of El Nino

COMMODITIES

Hurricane Nora is a warning of stranger upheavals to come in global weather

'El Meaño' responsible for disasters, benefits in world's weather

Predictions for the 1997-98 Cool Season

FIGURE 1-1.   A series of headlines illustrating the types of issues raised in the print media during the early phase of El Niño 97–98. These are from national newspapers and the Internet, as issued between September and November 1997.

media focus was on the bad outcomes expected from El Niño's influence in creating severe weather.

By November, El Niño had become part of the nation's popular culture. El Niño became the scapegoat for all types of problems totally unrelated to weather. Tornadoes occurred in California and Texas, and National Weather Service scientists attributed these to El Niño. Numerous commercial advertisements appeared in newspapers and on radio and television programs, urging purchases of products to thwart problems expected to result from El Niño, such as too much snow, crop losses and attendant economic market fluctuations, and flooding.

> *November 10 Advisory*—Warmest ever SSTs for October in parts of Pacific Ocean. New prediction of El Niño's duration—now expected to last through February–April 1998. Predictions continued to be for wet conditions in California and in the southern states and for warm weather in the northern states

Information continued to be widely released predicting detrimental future effects of El Niño on the economy and human health. Some meteorologists expressed concern over the fact that the public and business community had developed major misconceptions about El Nino's influence on the weather and the economy. However, sampling of public attitudes and many weather-sensitive decision makers in the Midwest and the Northeast revealed that these officials had largely ignored the winter climate predictions in their regions.

December 1997–January 1998

The first coastal storm to be blamed on El Niño hit California early in December, and El Niño's influence also was blamed for a bad Texas flood-producing rainstorm. The number of statements and proclamations about the dire consequences of the impending El Niño weather conditions continued to increase. The public began to believe the El Niño–based climate forecasts, and many people made plans to adjust to the expected winter weather.

> *December 11 Advisory*—SST anomalies increased during November, reaching the highest levels ever recorded. Prediction that ENSO would last at least through spring (March–May).

The peak of warming in the Pacific occurred in December, with SSTs of 28° to 29°C filling the equatorial basin. The first long-range El Niño–based experimental predictions of late spring and summer climate conditions were issued in December. Two scientific papers were published, both of which indicated that El Niño conditions were not tied to global warming.

Many new predictions relating to future conditions in the tropical Pacific were issued in January by various institutions. Most predictions called for El Niño's influence to strengthen and persist through the spring.

January 12 Advisory—Continued warm SSTs with "unprecedented" high values in the fall season. The "anomalous" convective pattern over the Pacific had lasted since May 1997. *Prediction 1*: The warm episode in the Pacific to continue through April–June 1998, then weaken. *Prediction 2*: Increased storminess to occur in California and along the southern tier of states.

News stories were almost devoid of scientific sources. Journalists, many of whom were not science writers, identified El Niño as the cause of storms of all types, often without consulting scientists. Bad storms occurred on both the west and east coasts, and all were attributed to El Niño. A very severe ice storm struck the northeastern United States and Canada early in January, causing $400 million in losses in the United States ($5 billion in Canada), and some scientists blamed the event on El Niño.

February–March 1998

Many scientific claims and news stories were published touting the accurate El Niño–based winter (December–February) predictions. These months were labeled "the year without a winter" in the nation's north (see Figure 1-2). The dramatic fall in energy costs became a major topic of conversation in the northern United States. Predictions continued about El Niño's societal impacts in the coming months, including its effects on jobs, health, agriculture, and social activities. Major scientific differences were revealed in the climate predictions issued about the climate conditions for the coming summer and beyond, and some scientists predicted the onset of La Niña conditions by the summer of 1998.

February 11 Advisory—The SST in the eastern Pacific increased during January 1998, and the warm episode was predicted to continue through spring. The El Niño circulation pattern over North America in February was predicted to continue into March.

Numerous forecasts about the 1998 crop season weather conditions and the crop yield effects appeared, and they were often in total disagreement.

Claims that El Niño 97–98 was the worst ever continued to appear in February. A tornado outbreak struck Florida, killing forty-two persons, and scientific debates over whether these and other winter tornadoes were caused by El Niño escalated. The news media published and broadcast a plethora of stories about El Niño's impacts, both good and bad, showing that there were major regional differences, all attributed to El Niño. Damaging storms occurred on both coasts, and the National Weather Service staff members blamed them on El Niño.

More predictions appeared in March calling for drastically different outcomes for the summer and fall of 1998. A sudden weather change with major winter storms and cold air from Canada swept across the central United States during

Crops may survive El Nino

'Marginal' price effect is expected

Let's Blame it on El Niño!

Rescuers keep searching

■ Hope fades for finding more survivors of Florida tornadoes

NATIONLINE

Texas braces for second round of rain and snow

More bad weather is expected today in Texas, where a weekend storm pummeled wide sections of central, northern and eastern parts of the state with up to 8 inches of rain and snow. Streets and rivers were flooded, and eight t...

El Nino-charged rain pounds California

LOS ANGELES (AP) — The season's most potent El Nino-charged storm took two lives ... have been evacuated around the lake.

The fight to save homes was...

The year without a winter

Slide in Energy Prices May Not Be Done

Energy Futures Fight a Losing Battle

Front-month contracts, daily data

COMMODITIES

By TERZAH EWING
Staff Reporter of THE WALL STREET JOURNAL
Black gold is tarnished, and natural ... is deflating ... ... in energy markets
... devoil

Crude-Oil Futures
Dollars per barrel
$20.0

El Nino Volume And Area Decrease, But Weather-Altering Power Still There

The 1997 El Niño/Southern Oscillation (ENSO 97-98)

– one of the most severe ENSO events in history?

FIGURE 1-2.   Headlines illustrating some of the key issues related to El Niño during its later stages, as selected from national newspapers published between January and May 1998.

the first half of March and temporarily ended the El Niño type of weather pattern throughout the country.

> *March 10 Advisory*—Very warm SSTs continued through February and altered the jet stream over North America, creating the predicted wet conditions in California, throughout the Gulf states, and in Florida. *Prediction 1*: The warm episode would continue through May. *Prediction 2*: The ENSO-related circulation pattern would continue through March.

NOAA officials announced that January–February 1998 conditions in the United States were the warmest and wettest ever and verified that CPC's winter weather forecast had been correct for most parts of the country. Figure 1-3 shows the winter weather outcomes for various regions across the nation. Global climate change as a reason for the record size of El Niño 97–98 appeared in the press again, and the cause-and-effect relationship continued to be debated by scientists. NOAA leaders observed that the El Niño winter weather conditions were similar to those expected under a global warming.

April–June 1998

> *April 10 Advisory*—Very warm ENSO conditions continued in Pacific, but SSTs were weakening, although the weakening was due to the normal annual cycle. The overall tropical circulation pattern continued unabated since June 1997. *Prediction 1*: Climate models called for continued warm conditions through April–June, then a return to "normal" during July–September. *Prediction 2*: Recurrent periods of significant storm activity and precipitation would occur across California and in the southern tier of states during April.

Predictions issued about climate conditions for the summer and fall of 1998 continued to differ greatly. News stories revealed that many news writers had gone back to the use of scientists as key sources, likely reflecting the new uncertainty surrounding the wide differences in predictions relating to El Niño and La Niña.

Several scientists, including those in the National Weather Service, and various media outlets linked a rash of tornadoes and tornado deaths in the Deep South to storms created by El Niño. Debates continued among scientists about what weather events to blame on El Niño. Forecasts issued in April predicted summer outbreaks of pests and both good and bad crop yield effects, as illustrated in Figure 1-2. Heavy rainfall across the Midwest and ensuing local floods were blamed on El Niño. The resulting wetness delayed crop planting in the Midwest, whereas unexpected severe droughts developed in Texas and Florida, the opposite of the wet weather that had been predicted to occur in those areas.

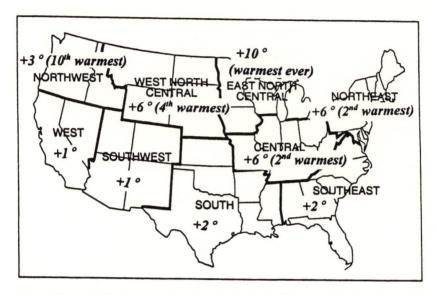

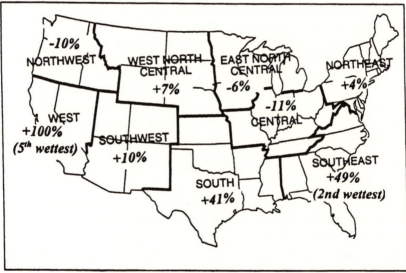

FIGURE 1-3. The winter (December 1997–February 1998) temperatures were much above normal in the northern sections of the nation (top map). The same season was rated as the warmest in the past 103 years in the upper Midwest and as the second warmest in the lower Midwest and the Northeast. The winter precipitation (lower map) was much above normal in the West, South, and Southeast regions of the nation. However, it was below normal in the two regions of the Midwest, and near normal elsewhere in the northern United States.

May 11 Advisory—Strong SST conditions persisted in the tropical Pacific but continued to decrease in magnitude during April. One forecast model indicated a return to near normal conditions in the Pacific during the next three to six months, whereas another climate model indicated more rapid cooling during the summer, with cooler than normal conditions likely to prevail during the second half of 1998.

The trade winds over the eastern Pacific abruptly returned to normal strength in mid-May, allowing cold subsurface waters to upwell, and the SSTs plummeted. Thus, El Niño was brought to an end, and La Niña (cold) conditions were established. This sudden ending was not anticipated in any of the ENSO forecasts issued during the spring.

Widespread forest fires in Mexico and in other parts of Central America during May and June produced enormous smoke plumes that entered the United States, and these fires were blamed on droughts caused by El Niño. Bad fires erupted in Florida after a very dry spring, and they were linked to El Niño, on the basis of claims that lush undergrowth had developed during the state's extremely wet winter, a condition predicted by the 1997 El Niño forecasts.

Speculation continued about the 1998 growing season conditions, since the long-range predictions diverged. In June, NOAA scientists declared El Niño was over because of the rapid cooling occurring in the Pacific. El Niño, which had been a major media topic since August 1997, largely disappeared from the news agenda.

June 9 Advisory—Warm ENSO conditions rapidly dissipated during May. Forecasts indicated cold episode conditions in the tropical Pacific during the rest of 1998, and all models agreed that a cold episode would develop during the coming six months and would last through the winter of 1998–1999.

Chapter 2 first presents a description of ENSO process and how El Niño conditions affect the weather conditions across North America and then examines the official climate predictions and their accuracy. Further information on the temporal distribution of scientific issues arising during El Niño is presented in chapter 4.

THE MAJOR EFFECTS: WINNERS AND LOSERS

When the El Niño event developed in 1997, most ENSO experts expected the resulting weather conditions to have a serious if not a disastrous impact on human lives and on the U.S. economy. The forecasting parallel was the strong 1982–1983 El Niño. Several scientists and institutions noted that the estimated global losses attributable to that event had ranged from $8 billion in losses and 1,500 deaths (*St. Louis Post-Dispatch*, June 18, 1997) to $13.6 billion with 2,000 deaths (*Cleveland Plain Dealer*, August 24, 1997). Hence, many prognostications issued in the summer and fall of 1997 were full of gloom and doom.

However, as El Niño 97–98 ended, the unexpected benefits accruing from the mild, dry, and nearly snow-free winter in the Midwest, Northeast, and High Plains, as shown in Figure 1-3, became recognized (*Detroit News*, March 20, 1998). For example, the unusual winter warmth led to estimated savings of $6.7 billion in heating costs to consumers, and the paucity of winter storms led to major reductions in spending on clearance of streets and highways, saving $21 million in the Chicago metropolitan area alone. An early estimate of losses and gains indicated that El Niño's net effect on the national economy was quite positive, with roughly $15 billion in gains as compared to $2 billion in losses (Michaels, 1998). Assessments early in 1999 suggested that the nationwide benefits accruing from El Niño included the saving of at least 800 lives and financial gains of nearly $19 billion. As Figure 1-4 illustrates, national retail sales set new records each month during the record warmth of January–April 1998. This was also a time when the U.S. economy was robust.

El Niño 97–98 had its share of storms, heavy rainfall, and major losses, too. Winter storms produced heavy precipitation in the West and the Southeast, causing widespread flooding. Federal and state government spending for relief and assistance for storm problems in California was $370 million, and in Florida it

FIGURE 1-4.   Record high winter temperatures in the Midwest and the Northeast brought people out of doors, and many went shopping in months when they normally did little shopping. Here, in February 1998, thousands of shoppers crowd downtown Chicago, helping to set all-time records for retail sales in the United States during the winter of 1998.

totaled $250 million. A fierce winter storm in January 1998 caused $400 million in losses and killed twenty-eight people, and total damages in Florida attributed to El Niño storms were $500 million (Ross et al., 1998). By late March 1998, FEMA had committed $289 million for relief, about the same as the $294 million spent in 1996–1997 and the $280 million in 1995–1996, both of which were not El Niño winters.

Fifteen catastrophes, events each causing more than $25 million in insured property losses, occurred between October 1997 and May 1998, resulting in losses of $1.7 billion nationally. The El Niño–related catastrophes were fewer and less costly than the twenty catastrophes and $2.1 billion (adjusted to 1998 dollars) in insured losses experienced in El Niño 1982–1983. Near-record heavy winter precipitation fell in California and in the Southeast, as shown in Figure 1-3.

The 1982–1983 El Niño created U.S. losses in excess of $2.2 billion (in 1983 dollars), or $4.6 billion when converted to 1998 dollars (Glantz, 1996). The warm winter of 1982–1983 in the Northeast, the warmest in twenty-five years, led to estimated savings of $500 million in heating costs. Deaths caused by storms attributed to El Niño during 1997–1998 amounted to 189, as compared to the 160 lives lost due to the 1982–1983 event. The estimated losses from El Niño 1997–1998 were $4.2 billion, slightly less than those incurred in 1982–1983. Importantly, the 1997–1998 losses did not set new records for damages, as had been predicted when the record-size El Niño developed in the summer of 1997. The mitigation activities preceding the 1997–1998 event reduced losses, particularly in California. Nevertheless, flooding was a major California problem, as illustrated by the photograph in Figure 1-5.

The largest surprise about the societal and economic impacts of El Niño 97–98 was the fact that the estimated gains outweighed the estimated losses ($19 billion as opposed to $4 billion in losses) and that the losses were much less than were expected to occur in the United States. Gains and losses varied regionally, as illustrated by comparing activities in the Midwest (Figure 1-4) with those in California (Figure 1-5). El Niño was correctly labeled a "meteorological Robin Hood" because of the benefits it produced in one region (the nation's northern sectors) and the losses it caused in the South and on the West Coast (*Wall Street Journal*, April 17, 1998). An extensive analysis of the impacts resulting from the weather events attributed to El Niño is presented in chapter 6.

Did El Niño and its predictable effect on the nation's weather teach us to change the way we think about the weather? Do the public and many decision makers understand better how climate works? If the answer to either question is "yes," the event was a major benefit to science and society.

ENORMOUS MEDIA ATTENTION

The thirteen-month period during which El Niño developed, existed, and ended (May 1997–June 1998) brought forth an interesting set of events at the local,

FIGURE 1-5.   Heavy rains in California brought major flooding. Here a resident of Lakeport paddles down a flooded street in February 1998. This flooding resulted when the nearby Clear Lake reached the highest levels in 90 years. (Courtesy Robert A. Eplett, California Office of Emergency Services)

state, regional, and national levels. People throughout the country got involved, from the White House to the ordinary citizen. El Niño became a household topic, widely discussed, often misunderstood, and commonly used as an excuse for any type of problem. The president went to disaster areas created by El Niño-spawned storms, and in October 1997 the vice president attended a mitigation conference in California organized to prepare for expected damaging storms.

The news media made El Niño a household word, as explained in chapter 3. By October 1997, practically every issue of every newspaper carried a story referring to El Niño, and evening television shows featured special presentations on the phenomenon. El Niño became a part of the popular culture and often became an excuse for human failure and the explanation for every conceivable problem, as illustrated in Figure 1-6.

The enormous media attention to the subject is partly reflected in the number of stories issued during El Niño 97–98. Table 1-1, based on a comprehensive survey of eight major Midwestern newspapers, presents a listing of the number of stories mentioning El Niño. This analysis of newspaper attention to the issue includes articles containing any mention of El Niño. The newspapers sampled included the *Chicago Tribune, Cincinnati Inquirer, Cleveland Plain Dealer, Des Moines Register, Detroit News , Indianapolis Star, Minneapolis Star Tribune*, and

FIGURE 1-6. The forecasts of bad weather resulting from El Niño issued in the summer 1997 brought enormous media coverage to the event. By the fall of 1997, "El Niño" had become a household word and was given as the reason for almost any conceivable failure or problem. This syndicated Sunday cartoon strip, from February 1998, illustrates the fame El Niño achieved as the cause for anything and everything. (Copyright Tribune Media Services, Inc., all rights reserved, reprinted with permission)

TABLE 1-1. Frequency of news articles appearing in eight major Midwestern newspapers containing references to El Niño, July 1997–May 1998.

	July	Aug.	Sep.	Oct.	Nov.	Dec.	Jan.	Feb.	Mar.	Apr.	May
No. of days with articles	3	8	13	23	19	22	21	27	28	16	7
No. of articles	3	11	21	42	32	43	44	114	112	56	21

St. Louis Post-Dispatch. The number of newspaper articles containing references to El Niño information totaled 502 for the July 1997–May 1998 period. The *Los Angeles Times*, which presented more El Niño-related coverage than any other major U.S. newspaper, had a total of 964 El Niño articles in the twelve-month period ending in May 1998.

The number of newspaper articles showed dramatic month-to-month fluctuations. Articles appeared on only eight days in August, but by October there were twenty-three days with articles appearing in one or more of the eight Midwestern newspapers, a level sustained for the next three months. The frequency increased again in February, with articles appearing almost every day, and with four or five of the eight newspapers carrying one or more articles each day. Analysis of major newspapers published on the West Coast and on the East Coast also shows that the peak month for news stories was February. However, it should be noted that by January 1998, many of the "El Niño" articles in the Midwest carried only a casual reference, such as "the local football game was canceled because of nasty El Niño weather," or perhaps some form of bad luck had occurred locally and was jokingly identified as a result of El Nino. By the fall of 1997, El Niño had become a household word signifying an excuse for failure or the cause of any type of problem. The topic had invaded the nation's popular culture. Newspaper cartoons were filled with El Niño jokes, as shown in Figure 1-6, and numerous advertisers used El Niño as a reason for consumers to buy their products.

Analysis of the number of El Niño stories appearing each month on the Internet between June 1997 and August 1998 also shows a peak in the frequency of information during February–March 1998. A good statistical relationship was found between the temporal distribution of Internet articles and the Southern Oscillation Index (SOI), a scientific measure of the strength of the El Niño conditions in the Pacific Ocean—the SOI peaked in January–March 1998 (Hare, 1998). However, as shown in chapter 3, the peaks in the number of news stories coincided with the occurrence of severe weather events during this late-winter period, and the statistically significant relationship to the SOI is likely a coincidence.

REACTIONS TO EL NIÑO

The evolution of the event, from its rapid springtime development to its record proportions during the summer–fall of 1997 and its dissipation in May–June

1998, brought both disastrous and beneficial weather conditions to the United States. An unpredictably diverse number of reactions to El Niño occurred across the nation at all levels, from the public to weather-sensitive decision makers to government officials.

The initial reactions were related to the seasonal climate forecasts issued during June–August 1997, all carefully couched in scientific terms. These predictions brought mixed responses, with some decision makers deciding to act and many choosing to ignore the forecast conditions. The most common reaction from the public and from decision makers was to ignore the predictions, although their applications are described in the next section. Public surveys done in October revealed that most people doubted the predictions' accuracy, and many said they were confused by the El Niño phenomenon and how it could affect the weather, as illustrated by the cartoon presented in Figure 1-7.

Experts in a variety of fields were asked to speculate on or predict—or proclaimed their beliefs on their own about—what would happen if the predicted climate conditions were actually to occur. Their speculations ranged widely, and once the weather of 1997–1998 had occurred and the actual impacts had been realized, many of them turned out to be erroneous. As is shown in chapter 6, diverging expert opinion revealed a lack of understanding of complex weather impacts.

Events early in the fall, including the lack of Atlantic hurricanes and an October snowstorm in the High Plains, both attributed to El Niño by scientists, indicated a shift in perceptions about El Niño's influence on the nation's weather. Science-based articles were overwhelmed by articles attributing every weather event and its outcome to El Niño. In the popular culture, El Nino had become accepted as a climatological fact.

FIGURE 1-7.   One of the interesting reactions of weather-sensitive decision makers and the general public to the long-range climate forecasts calling for bad weather as a result of El Niño was uncertainty over what to do. This cartoon strip, from a Denver newspaper, aptly illustrates that, although there was historical evidence and strong science behind the forecasts, there were still major uncertainties about whether or how to react. (Reprinted with permission of Ed Stein, courtesy of the *Rocky Mountain News*)

However, in California, where El Niño 1982–1983 had caused major losses, the public and state decision makers paid attention. Individuals reacted and spent an estimated $125 million on home and business repairs and improvements to mitigate the effects of the predicted heavy rains, high winds, and flooding.

Everywhere the public reacted differently as the winter weather unfolded. Those residing in the northern sectors of the nation, where near record warmth existed, went outdoors, fished, played golf, or went walking, hiking, and shopping. Vacation plans were shifted as skiers without snow in the Midwest flew west to the snow-covered mountains of California and Utah, and trips to Florida and California coastal resorts, where the weather was unusually wet and stormy, were canceled. Northern restaurants did land-office business, and national home sales set all-time records for the January–April period, a time of expected downturn even when the economy is good. The warmth reduced heating bills by 10 to 30 percent, leading to major savings for millions of individuals and businesses.

Those unfortunate enough to experience the bad weather were in the southern United States or in California. Those with damaged homes and businesses sought government relief, and insurers paid. California received $370 million in federal and state aid. Many farmers with flooded fields in California and in Florida raised prices for their fresh vegetables and got government relief. A massive tornado outbreak killed forty-two persons in Florida, a state that normally has few tornadoes. The event illustrated the dangers of El Niño weather.

Atmospheric and oceanic scientists also reacted to El Niño. The event led to new levels of research about the phenomenon. The number of scientific papers on El Niño presented at national conferences doubled between 1997 and 1998, as described in chapter 4. There was new interest in, and new government funds made available for, the study of El Niño. Further information about the public's reactions to El Niño is presented in chapters 3 and 6, and chapter 7 presents a series of interesting case studies illustrating how policy makers reacted to El Niño.

The federal government and its agencies reacted in interesting ways. NOAA, the official weather forecasting agency, held agency and press briefings describing its climate predictions. FEMA took these forecasts and issued warnings of potential flooding problems to generate interest in mitigative activities. Other agencies modified their field operations for expectations of above-normal rainfall and helped perform mitigative activities. Congress held hearings to ascertain what was being done. After the weather damages occurred, FEMA and other agencies stepped in and gave assistance and funds for relief. There was also an international incident. El Niño had produced the worst drought in seventy years in parts of Mexico. By late May 1998, there had been 11,000 fires, burning more than one million acres, killing fifty Mexicans, and creating huge smoke palls that spread into the southern United States, creating severe air quality problems as the smoke moved as far north as Chicago (*Chicago Tribune*, May 26, 1998). The Mexican government initially refused any U.S. aid, including an offer of $5 million, but after several tense negotiations, Mexico relented and allowed U.S. firefighting experts to help deal with the fires.

APPLICATIONS OF THE EL NIÑO PREDICTIONS: USERS ARE WINNERS

A key issue integral to El Niño 97–98 was the accuracy of long-range climate forecasts based on knowledge about El Niño's influence on the atmosphere. Once the oceanic warming of El Niño began developing rapidly in the tropical Pacific during May 1997, atmospheric scientists at NOAA and other research institutions began predicting El Niño's effects on climate conditions around the world. By late June, the official U.S. climate outlooks for the fall, winter, and spring seasons called for wet and stormy conditions in California and in the southern states, with warm and dry conditions expected in the northern two-thirds of the nation (see Chapter 5). As the weather unfolded during these three seasons, the predictions were found to be quite accurate. In 1999, the director of the official forecast group in NOAA announced that past research and oceanic data had "led to an incredibly bold forecast of El Niño nearly six months prior to the onset of the major impacts" (Leetma, 1999). The successful forecasts served as an enormous boon to the science of long-range climate prediction and proved the value of the investment in oceanic and atmospheric research made over the preceding twenty years.

Scientists and journalists used several fall weather events as a verification of the forecasts (see chapter 4). El Niño quickly became a "fact," as reported in the media and, in turn, believed by the public. After this nationwide acceptance developed, there was widespread belief in the subsequent predictions about winter and spring conditions. Public polls showed that many lay people had reversed their skeptical views about the impact of El Niño. For example, as previously noted, many consumers changed their winter vacation plans, and home buying rose to record levels. In general, the highly accurate predictions served to impress many and gave the public a new image of science's ability to forecast the weather months in advance.

In addition to the media and the public, the private sector, government, and the scientific community also made major use of the predictions. The first institutions to put the predictions to use were government agencies. Federal agencies, beginning with NOAA, which issued the forecasts and held agency briefings, and also including FEMA and the Army Corps of Engineers, made plans to mitigate damages. The Geological Survey and the Bureau of Reclamation moved to more effectively manage western water supplies in rivers and reservoirs, and the Environmental Protection Agency acted to minimize environmental damage. Hearings were held to brief members of Congress, and by early September, the federal government was moving along several fronts.

State and regional agencies also reacted to the predictions, partly in response to the federal warnings. Regional climate centers issued region-specific outlooks addressing weather conditions not included in the official forecasts, such as the prospects for winter snowfall in the Midwest. State agencies in California and Florida reacted to NOAA's predictions and to FEMA's warnings and launched mitigation efforts aimed at reducing storm damages and flood losses. California communities also reacted with local flood mitigation efforts, and some Midwestern agencies held off on buying salt for roads and highways.

These governmental actions have not been assessed for their economic impacts, but they appear to have produced sizable benefits. For example, in California, where storms and heavy rains caused by El Niño 1982–1983 created devastating losses, the state government, communities, and many citizens reacted quickly. In August and September, with prompting from FEMA, the state and many communities launched major mitigative efforts to prepare for the predicted storms, heavy rains, and floods. The costs, both personal and government related, were assessed at $165 million. The 1997–1998 storm losses in California caused $1.1 billion in damages, but this was half the $2.2 billion (1998 dollars) in losses caused in California by the comparable 1982–1983 event, suggesting that the $165 million expenditure resulted in a sizable reduction in losses.

The economic values associated with the use of these seasonal predictions were assessed for certain businesses (see chapter 5). Prior studies of the use of long-range predictions in agribusiness and power utilities revealed that only 25 percent to 30 percent of the potential users in weather-sensitive positions actually used long-range forecasts in their decisions (Changnon et al., 1995). Study of the actions of eighty-seven decision makers during the 1997–1998 event revealed that 47 percent of those sampled (50 percent more than in prior years) had chosen to utilize the 1997–1998 predictions, a marked upward shift. The greater accuracy in the 1997–1998 predictions offered an opportunity to react and to take into account the predicted conditions in making both operational and planning decisions. Utility users employed the predictions to schedule various weather-sensitive activities and to trade and sell natural gas and power and as a result realized benefits to individual firms ranging from a low of $200,000 to as much as $30 million. Some utilities that provide natural gas reacted to the above-normal temperature forecasts for the northern United States and held off on making early-season commitments to buy their winter supplies. As gas prices dropped precipitously during the warm winter, they bought from the market and saved millions of dollars. Three Midwestern companies that sell natural gas and that used the predictions to guide their decisions saved their customers $39 million, $43 million, and $82 million, respectively (see chapter 6).

Elsewhere in the nation, reactions to the predicted conditions were highly varied. Commodity brokers and certain businesses used the predictions of bad weather to launch major promotional sales programs advertising and selling their services and products. Non-use of the El Niño predictions was most common in agribusinesses (the forecasts were not applicable to the 1998 growing season) and among several in government positions responsible for sensitive activities involving water management, fuel acquisition, and highway clearance (for winter conditions). These decision makers indicated they were afraid to take the risk of using the forecasts to alter their normal operations. Many private-sector weather forecasting companies used the forecasts, often modifying and interpreting them to meet their clients' needs. Chapter 7 addresses the issue of use and misuse of the El Niño climate forecasts.

Atmospheric scientists also used the predictions. Some challenged the accuracy of the forecasts, and others analyzed the forecasts and launched new research. Some scientists used the forecasts to develop their own predictions for local areas or for special weather conditions, such as the number of winter storms on the Great Lakes. In one such case, described in chapter 5, one Midwestern university that relied on a forecast promoted by one of its professors saved $500,000 on its winter heating costs. Several scientists were overwhelmed by the success of their forecasts, and, as one El Niño forecaster noted, "We now have a scientific basis for climate prediction, and that suggests that the larger-scale effects for all future major El Niño events should be predicted several months in advance" (Shukla, 1998).

EL NIÑO AND GLOBAL WARMING

During the thirteen-month period of El Niño, several contentious scientific issues emerged. Controversies included concerns about the accuracy of the long-range climate predictions, which weather events to attribute to El Niño, the accuracy of widely diverse predicted climate effects, and the causes of the record-size El Niño. However, the most volatile of the scientific issues concerned the relationship between the record-setting size of El Niño 97–98 and global warming. This issue actually includes two intertwined questions:

1. Was El Niño caused or enhanced by global warming?
2. Was the El Niño winter weather an indicator of what global warming would be like?

Many scientists labeled El Niño 97–98 the "climate event of the century." The media publicized the scientific claims, which, in turn, raised questions about causality. One explanation was that El Niño's strength was a product of global warming. Monthly temperatures from December 1997 through May 1998 were well above normal, as illustrated for December in Figure 1-8, and this helped create a belief that there was a relationship between global warming and El Niño.

A question that emerged during the event is whether El Niño 97–98 and its much-above-normal temperatures (see Figure 1-3) were a harbinger of the cold-season weather conditions that might be expected under global warming. NOAA Administrator Baker called the near-record warm winter of 1997–1998 a possible "window on the future if overall global warming projections pan out" (*USA Today*,

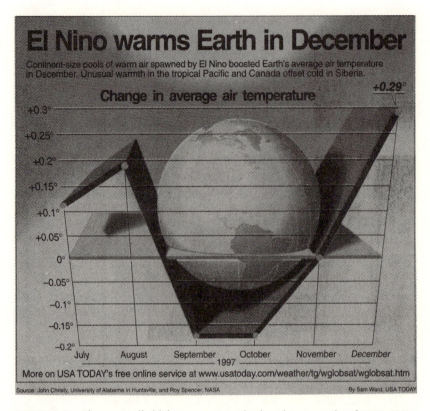

El Nino warms Earth in December

Continent-size pools of warm air spawned by El Nino boosted Earth's average air temperature in December. Unusual warmth in the tropical Pacific and Canada offset cold in Siberia.

Change in average air temperature

+0.29°

+0.3°
+0.25°
+0.2°
+0.15°
+0.1°
+0.05°
0°
−0.05°
−0.1°
−0.15°
−0.2°

July   August   September   October   November   December

1997

More on USA TODAY's free online service at www.usatoday.com/weather/tg/wglobsat/wglobsat.htm

Source: John Christy, University of Alabama in Huntsville, and Roy Spencer, NASA

By Sam Ward, USA TODAY

FIGURE 1-8. The unusually high temperatures in the winter months of 1997–1998 attracted the attention of atmospheric scientists and the press. This graph illustrates how the scientists and press tied the high global temperatures of December 1997 to El Niño's influence on global circulation. (Copyright 1998, *USA TODAY*, reprinted with permission)

April 7, 1998). A climatologist at the National Center for Atmospheric Research claimed that there was an excellent analogy between El Niño 97–98 and the kinds of effects people will experience under climate change due to global warming (*Climate Alert*, 1998). However, some journalists noted that the great benefits accruing from El Niño, when compared with climatologists' early predictions of great weather-induced losses for 1997–1998, raised serious questions about the accuracy of climatologists' predictions about global warming and its problems.

The second question is whether El Niño 97–98 was a *product* of global warming. At a California mitigation workshop in October 1997, Vice President Al Gore expressed the view that global warming was the cause for El Niño 97–98, but some journalists questioned whether the assertion was just another way to promote public concern over global warming. Several scientists published papers indicating that there was no tie between global warming and El Niño 97–98, whereas others published papers claiming a physical relationship existed; this debate was discussed in some news accounts. A White House official noted in June 1998 that Vice President Gore, after announcing to the national media that the record high U.S. temperatures from February to May were proof of global warming, was linking El Niño and global warming to further his goal of getting a $6.3 million program to reduce atmospheric emissions through Congress (*Minneapolis Star Tribune,* June 8, 1998).

Chapter 3 explores the possibility that El Niño 97–98 was a "signal event," an event, as defined in risk communication, that sends a political and cultural signal about a high-risk issue that is much more significant than the outcome of the actual event. Did the media coverage and input from several scientists (tying global warming to El Niño) help create a signal event about the possible impacts (or reality) of global climate change? Karl (1998) noted that "being able to forecast a season ahead will help us if our worst fears about more extreme weather (due to global warming) come true." Furthermore, accurate scientific predictions about the fall and winter weather may have helped solidify public and political belief in such anomalous climatic events.

Yet, there are reasons why El Niño 97–98 may not become a signal event. El Niño weather damages were vividly displayed on TV, and this helped create the image of "strange weather." However, as Ungar (1999) explains, this does not necessarily convert into an awareness of climate change. Second, the net effect of El Niño on the United States was positive; the losses of life and dollars were much lower than predicted and the benefits in lives saved and the economic gains outweighed the losses. Hence, the event was far from the expected disaster predicted by many in 1997. Further, the weather predictions issued for the seasons after the winter of 1997–1998 were widely divergent. Few forecasters exhibited any skill in predicting the climate conditions of the summer of 1998 and the seasons that followed. Many users of the El Niño predictions interviewed felt that the successful El Niño predictions were anomalous and expected a return to average forecasting accuracy (chapter 5). Thus, the impact of the successful predictions may be forgotten in the coming years.

El Niño 97–98 can be viewed as a boost for both sides of the scientific-political debate over what should be done about controlling emissions of greenhouse gases to halt or slow global warming. One side saw the strong El Niño as evidence of human meddling with the fragile earth, whereas the other side saw El Niño as nature asserting itself at the top of the earthly order (*Minneapolis Star Tribune,* October 29, 1997). It is too soon to determine whether El Niño 97–98 served as a signal event for global climate change. Chapters 3 and 4 present extensive information about the El Niño-global warming issue.

SCOPE OF THIS BOOK

This book has eight chapters. Chapter 2 addresses the oceanic and atmospheric conditions in the tropical Pacific that created El Niño 97–98, including how the phenomenon affected weather patterns over the United States; discusses how scientific predictions based on El Niño were generated; and presents the seasonal weather conditions that actually occurred in 1997–1998. Chapter 3 describes how the news media interpreted and translated the scientific information generated about El Niño and how the public reacted.

Chapter 4 assesses the scientific issues that emerged as a result of El Niño. Chapter 5 describes how decision makers in weather-sensitive positions in agribusiness, water management, and utilities reacted to the El Niño-based seasonal climate predictions, assesses the uses of the predictions, and identifies the reasons for their nonuse. Chapter 6 presents a national and regional analysis of the impacts of El Niño-caused weather conditions across the nation, defining the winners and losers.

Chapter 7 addresses inherent problems in using predictions and describes some of the key policy actions resulting from El Niño-based predictions and weather problems. Chapter 8 summarizes the key findings, including how decision makers and the public reacted to El Nino and predictions about it and identifies the major scientific surprises related to El Niño, the lessons learned from the event, and the legacy of the event for scientists and the nation.

Many of the illustrations presented in this volume were selected to emphasize how the media presented information about El Niño. Predictions, outlooks, and forecasts are terms used interchangeably throughout the text and are considered to have the same meaning in relation to future oceanic, weather, or climate conditions.

REFERENCES

Changnon, S. A., Changnon, J. M., and Changnon, D. 1995. Uses and applications of climate forecasts for power utilities. *Bulletin Amer. Meteoro. Soc.*, 76, 711–720.
Chicago Tribune. May 26, 1998. Mexico's fires cloud horizons, p. 1.
Climate Alert. July–August 1998. Is there a relationship between El Niño and climate change? vol. 12, p. 1.
Cleveland Plain Dealer. August 24, 1997. El Niño is making itself well known, p. 4H.
Detroit News. March 20, 1998. Farewell to the winter that wasn't, p. E1.
Glantz, M. H. 1996. *Currents of Change.* Cambridge: Cambridge University Press.
Hare, S. R. 1998. Recent El Niño brought downpour of media coverage. *EOS*, 34, 6.
Karl, T. April 7, 1998. El Niño could be window on future. USA *Today*, p. 1.
Leetma, A. 1999. The first El Niño observed and forecasted from start to finish. *Bulletin Amer. Meteoro. Soc.*, 80, 111–112.
Michaels, P. April 19, 1998. Beware, was El Niño combined with global warming? No! *Los Angeles Times*, p. 5A.
Minneapolis Star Tribune. October 29, 1997. A story of superlatives. p. 5A.
———. June 8, 1998. El Niño accelerates global warming trend, p. 3A.
Ross, T., Lott, N., McCown, S., and Quinn, D. 1998. *The El Niño Winter of '97–'98.* NCDC Technical Report No. 98-02, NOAA, Asheville, NC.

St. Louis Post-Dispatch. June 18, 1997. El Niño cooking up recipe for strange winter weather, p. 12A.

Shukla, J. 1998. El Niño and climate more predictable than previously thought. *Bulletin Amer. Meteoro. Soc.,* 79, 2816.

Ungar, S. 1999. Is strange weather in the air? A study of the U. S. national network news coverage of extreme weather events. *Climatic Change,* 41, 133–150.

USA Today. April 7, 1998. El Niño could be window on future, p. 1.

Wall Street Journal. April 17, 1998. Those corrections on page 5 are El Niño's fault. p. 1.

2

Causes, Predictions, and Outcomes of El Niño 1997–1998

VERNON E. KOUSKY &
GERALD D. BELL

WHAT IS CLIMATE VARIABILITY?

One of the most prominent aspects of our weather and climate is its variability. This variability ranges over many time and space scales, from small-scale weather phenomena such as wind gusts, localized thunderstorms, and tornadoes, to larger-scale weather features such as fronts and storms and to prolonged climate features such as droughts, floods, and fluctuations occurring on multiseasonal, multiyear, and multidecade time scales. Some examples of these longer time-scale fluctuations include abnormally hot and dry summers, abnormally cold and snowy winters, a series of abnormally mild or exceptionally severe winters, and even a mild winter followed by a severe winter. In general, the longer time-scale variations are often associated with changes in the atmospheric circulation that encompass areas far larger than a particular affected region. At times, these persistent circulation features affect vast parts of the globe, resulting in abnormal temperature and precipitation patterns in many areas. During the past several decades, scientists have discovered that important aspects of interannual variability in global weather patterns are linked to a naturally occurring phenomenon known as the El Niño/ Southern Oscillation (ENSO) cycle. The heart of ENSO lies in the tropical Pacific, where there is strong coupling between variations in ocean surface temperatures and the circulation of the overlying atmosphere. The terms El Niño and La Niña represent opposite extremes of the ENSO cycle, and they cause very different rainfall outcomes, as illustrated in Figure 2-1.

Before describing the oceanic and atmospheric characteristics of the ENSO cycle, it is necessary to describe the average climatic conditions and how they vary throughout the year.

FIGURE 2-1.   The different and quite extreme weather outcomes associated with the alternating El Niño and La Niña conditions in the Pacific Ocean are well illustrated by this roller coaster cartoon. (Reprinted with permission, Dana Summers, Orlando Sentinel)

Mean Oceanic and Atmospheric Conditions in the Tropical Pacific

Interannual climate variability is often measured by comparing the observed conditions to the long-term mean conditions. The mean state of the tropical Pacific Ocean is identified by both its surface and its subsurface characteristics, each of which exhibits considerable evolution across the eastern half of the tropical Pacific during the course of the year.

Throughout the year, the ocean surface is warmest in the west and coldest in the east. The largest difference between the two regions is observed during September and October, when temperatures in the eastern Pacific reach their annual minimum. Temperatures across the central and eastern tropical Pacific then normally begin to increase in November–December and peak in March–April. In contrast, sea surface temperatures(SSTs) across the western tropical Pacific and Indonesia remain warm and nearly constant throughout the year.

The ocean surface temperatures across the tropical Pacific contribute significantly to the pattern of deep tropical convection, associated with showers and thunderstorms. Throughout the year, the strongest convection and the heaviest rainfall are observed over Indonesia and the western tropical Pacific, and the weakest convection and the least rainfall are found over the eastern equatorial Pacific.

During the Northern Hemisphere cool season, the mean patterns of sea surface temperature and equatorial rainfall are accompanied by low-level easterly winds (east-to-west flow) and upper-level westerly winds across the tropical Pacific. Over the western tropical Pacific and Indonesia, this wind pattern is associated with low air pressure at the surface and rising air motion, while over the eastern Pacific it is accompanied by high air pressure at the surface and sinking motion. Collectively, these conditions reflect the equatorial Walker Circulation (Figure 2-2), which is a primary large-scale circulation feature over the Pacific during the northern cool season.

The low-level winds aid in the accumulation of warm water in the extreme western tropical Pacific and result in an upwelling of cold water along the equator in the central and eastern Pacific. This upwelling is strongest from June through September and weakest from December through April, and it is a major contributor to the annual cycle of SSTs just described. In general, the upper-ocean water is separated from the colder deep-ocean waters by the oceanic thermocline, which is normally deepest (150 m) in the west and shallowest in the east, where it is located 30 to 50 m below the ocean surface. The resulting east-west variations in upper-ocean temperatures result in east-west variations in sea-level height, which is higher in the west than in the east.

December - February Normal Conditions

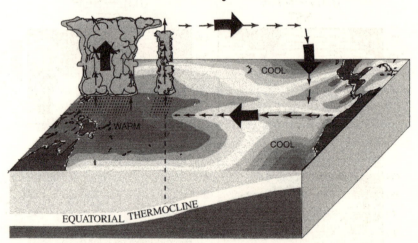

FIGURE 2-2. Schematic diagram of the normal east-west direct Walker circulation during December–February. Rising (sinking) motion is depicted over the warm (cool) waters of the western (eastern) tropical Pacific. Low-level easterlies and upper-level westerlies complete the circulation cell. The thermal structure of the upper ocean is also depicted in this figure. The warm upper ocean waters are separated from the deep cold ocean waters by a region of contrast, called the thermocline.

The Pacific Ocean temperatures, tropical rainfall and vertical motion patterns greatly affect the distribution of atmospheric heating across the tropical and subtropical Pacific. Normally, the strongest heating and the warmest air temperatures coincide with the warmest ocean waters and the heaviest rainfall. This atmospheric heating helps to determine the overall north-south temperature contrast in both hemispheres, which significantly affects the location of the jet streams over both the North and the South Pacific Oceans. This influence on the jet streams tends to be most pronounced during the respective hemisphere's winter season, when both the location and the eastward extent of the jets (to just east of the International Date Line—180° W) exhibit a strong relationship to the pattern of tropical heating (Figure 2-3). These jet streams are then a major factor controlling the winter weather patterns and storm tracks in the middle latitudes over the Pacific and immediately downstream over both North and South America.

What Is El Niño?

Near the end of each calendar year, ocean surface temperatures warm along the coasts of Ecuador and northern Peru. In the past, local residents referred to this annual warming as "El Niño," meaning The Child, due to its appearance around the Christmas season (Wyrtki, 1975). The appearance of El Niño signified the end of the fishing season and arrival of the time for Peruvian fishermen to repair their nets and maintain their boats.

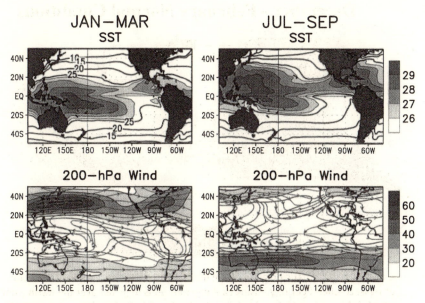

FIGURE 2-3. Climatological (long-term average) sea surface temperatures (C, top panels) and upper-tropospheric (200-hPa) winds (m s⁻¹, bottom panels) for January–March and July–September. The warmest temperatures and strongest wind speeds are indicated by dark shading.

Every two to seven years a much stronger warming appears along the west coast of South America, which lasts for several months and which is often accompanied by heavy rainfall in the arid coastal regions of Ecuador and Peru. Over time the term El Niño began to be used in reference to these major warm episodes. Beginning in the 1960s, scientists began to link the abnormally warm waters along the west coast of equatorial South America with abnormally warm waters throughout the central and east-central equatorial Pacific (Berlage, 1966). In addition, the warmer than normal waters were shown to be closely related to a global atmospheric air pressure oscillation known as the Southern Oscillation (SO) (Berlage, 1966; Bjerknes, 1966, 1969; Troup, 1965; Walker and Bliss, 1932).

The term El Niño now refers to the coupled ocean-atmosphere phenomenon characterized by (1) abnormally warm sea surface temperatures from the date line east to the South American coast (a distance spanning nearly one-fourth of the Earth's circumference along the equator), (2) changes in the distribution of tropical rainfall from the eastern Indian Ocean east to the tropical Atlantic (a distance spanning more than one-half the Earth's circumference), (3) changes in air pressure throughout the global tropics, and (4) large-scale atmospheric circulation changes in the tropics and portions of the extratropics in both hemispheres. Other terms commonly used for the El Niño phenomenon include "Pacific warm episode" and "El Niño/ Southern Oscillation (ENSO) episode."

Characteristics of El Niño

El Niño episodes are characterized by lower-than-normal air pressure over the eastern tropical Pacific and higher-than-normal air pressure over Indonesia and northern Australia. This pressure pattern reflects the negative phase of the SO and is associated with weaker-than-normal low-level equatorial easterly winds over the central and eastern Pacific. These conditions result in weaker-than-normal oceanic upwelling and warmer-than-normal sea surface temperatures in the central and eastern Pacific. These increased sea surface temperatures then favor the development of atmospheric convection and rainfall anywhere temperatures exceed 28°C (82°F) (Gadgill et al., 1984). These features tend to be strongest at the time of year when the ocean surface temperatures are seasonally warmest (January–April). However, during very strong El Niño episodes they may be present by June–August and persist for the next nine to twelve months. During these periods the east-west temperature difference normally observed al   the equator is much weaker than normal, and the Walker Circulation is nonexistent (Figure 2-4).

El Niño episodes, the eastward extension of deep tropical convection pospheric heating to well east of the International Date Line con-'ft toward the equator and a pronounced eastward extension to

'98

December - February El Niño Conditions

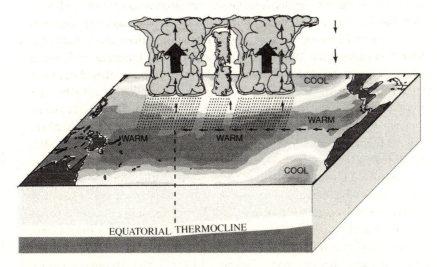

FIGURE 2-4.   Schematic diagram of equatorial Pacific oceanic and atmospheric El Niño–related features during December–February.

the extreme eastern Pacific of the mid-latitude wintertime jet streams in both hemispheres (e.g., Arkin, 1982). Overall, these conditions reflect a more zonally uniform distribution of both temperature and winds across the Pacific basin than is evident in the climatological mean, and they are a major factor affecting the winter weather patterns and storm tracks in the middle latitudes over both North and South America.

Variations in the strength and distribution of anomalous sea surface temperatures from one El Niño to another leads to variations in the extra-tropical response. During weak El Niño episodes the enhanced convection is confined to the vicinity of the date line, while during strong episodes it extends across the entire eastern half of the Pacific basin and covers a distance of approximately one-quarter of the Earth's circumference. Thus, the mean wintertime jet tends to be extended farther east during strong episodes. These variations in the jet ultimately result in event-to-event variations in the downstream circulation features and in the patterns of temperature and rainfall over much of North America.

Nonetheless, by studying past warm (El Niño) episodes, scientists have discovered precipitation and temperature anomaly patterns that are highly consistent from one episode to another (see, e.g., the pioneering works of Mossman, 1924, and Walker, 1923, 1924, 1928a, 1928b; and the more recent works of Aceituno, 1988; Bhalme et al., 1983; Caviedes, 1973; Halpert and Ropelewski, 1992; Hastenrath and Heller, 1977; Kiladis and Diaz, 1989; Kousky et al., 1984; Rasmusson and Carpenter, 1983; and Ropelewski and Halpert, 1986)

Significant departures from normal for the Northern Hemisphere winter (December–February) and summer (June–August) seasons are shown in Figure 2-5. Within the Tropics, the eastward extension of rainfall and thunderstorm activity toward the central and eastern Pacific results in abnormally dry conditions over northeastern Australia, Indonesia, and the Philippines in both seasons. During the northern winter season, drier-than-normal conditions are also observed over southeastern Africa and northern South America. Wetter-than-normal conditions are observed along the West Coast of tropical South America and along the Gulf Coast of the United States. During the northern summer season, the Indian monsoon rainfall tends to be reduced, especially in northwestern India, where crops are sometimes adversely affected. Drier-than-normal conditions are also experienced over Mexico and portions of Central America, while wetter-than-normal conditions are observed over subtropical latitudes of South America (central Chile, southern Brazil, Uruguay, and northeastern Argentina).

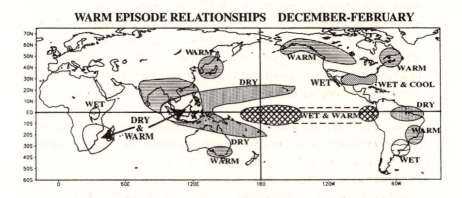

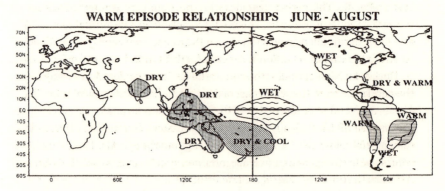

FIGURE 2-5. Typical Northern Hemisphere winter (December–February) and summer (June–August) temperature and precipitation anomalies associated with warm (El Niño) episodes.

During a warm episode winter, mid-latitude low pressure systems tend to be more vigorous than normal over the eastern Pacific and in the region of the Gulf of Alaska. These systems pump abnormally warm air into western Canada, Alaska, and the northern portion of the contiguous United States. Storms also tend to be more vigorous than normal in the Gulf of Mexico and along the southeast coast of the United States, resulting in wetter-than-normal conditions in those regions.

What Is the ENSO Cycle?

El Niño (warm) and La Niña (cold) episodes are extremes of what is often referred to as the El Niño/Southern Oscillation (ENSO) cycle. The cycle has an average period of about four years, although in the historical record the period has varied between two and seven years (see Table 2-1 for a list of El Niño and La Niña episodes since the late 1800s). The ENSO cycle encompasses opposite extremes in ocean surface and subsurface temperatures, tropical rainfall, atmospheric winds, and air pressure. During El Niño episodes, the equatorial sea surface temperatures (SSTs) are abnormally warm from the date line (180° W) east to the South American coast. However, there is generally a strong annual cycle in the actual SSTs, so the waters are sufficiently warm (approximately 28°C [82°F] (Gadgill et al., 1984) to support persistent tropical rainfall and convective activity across the eastern equatorial Pacific for only part of the year. Thus, this annual cycle strongly affects the timing and eastward extent of tropical rainfall during the El Niño.

As a typical El Niño develops, above-normal rainfall tends to extend east to just east of the date line during September–November. The lack of rainfall farther east coincides with the minimum in SSTs across the eastern equatorial Pacific at this time of the year. During December–January, tropical rainfall extends to well east of the date line, and by March–April it typically covers the entire eastern Pacific. This period coincides with the annual maximum in SSTs across the eastern equatorial Pacific. This disruption of the normal annual cycle of SSTs and tropical rainfall also acts to weaken the major monsoon systems affecting Australia/Southeast Asia, South America/Central America, and tropical Africa.

During La Niña episodes, the equatorial SSTs are colder than average from the date line east to the west coast of South America, and tropical rainfall and convection tend to be focused over the western equatorial Pacific and Indonesia. Little rainfall is typically evident over the eastern equatorial Pacific, as SSTs remain well below 28°C in this region throughout the episode. These conditions contribute to stronger-than-average monsoon systems over Australia/Southeast Asia, South America/Central America, and Africa.

The evolution of El Niño and La Niña episodes, as well as the transition between these extreme phases of the ENSO cycle, depends greatly on the subsurface ocean temperature structure and the variability of the low-level winds. As

TABLE 2-1. Pacific warm (El Niño) and cold (La Niña) episodes

Year	Episode	Intensity	Year	Episode	Intensity
1877–1878	Warm	Strong	1951	Warm	Weak
1886	Cold	Moderate	1953	Warm	Weak
1888–1889	Warm	Moderate	1954–1956	Cold	Strong
1889–1890	Cold	Strong	1957–1959	Warm	Strong
1896–1897	Warm	Strong	1963	Warm	Weak
1899	Warm	Weak	1964–1965	Cold	Moderate
1902–1903	Warm	Weak	1965–1966	Warm	Moderate
1903–1904	Cold	Strong	1968–1970	Warm	Moderate
1905–1906	Warm	Strong	1970–1971	Cold	Moderate
1906–1908	Cold	Strong	1972–1973	Warm	Strong
1909–1910	Cold	Strong	1973–1976	Cold	Strong
1911–1912	Warm	Strong	1976–1977	Warm	Weak
1913–1914	Warm	Moderate	1977–1978	Warm	Weak
1916–1918	Cold	Strong	1979–1980	Warm	Weak
1918–1919	Warm	Strong	1982–1983	Warm	Strong
1923	Warm	Moderate	1983–1984	Cold	Weak
1924–1925	Cold	Moderate	1984–1985	Cold	Weak
1925–1926	Warm	Strong	1986–1988	Warm	Moderate
1928–1929	Cold	Weak	1988–1989	Cold	Strong
1932	Warm	Moderate	1990–1993	Warm	Strong
1938–1939	Cold	Strong	1994–1995	Warm	Moderate
1939–1941	Warm	Strong	1995–1996	Cold	Weak
1946–1947	Warm	Moderate	1997–1998	Warm	Strong
1949–1951	Cold	Strong	1998–1999	Cold	Strong

Warm (ENSO) episodes sources: Rasmusson and Carpenter, 1983, *Monthly Weather Review*; Ropelewski and Halpert, 1987, *Monthly Weather Review*. Cold Episode sources Ropelewski and Halpert, 1989, *Journal of Climate; Climate Diagnostics Bulletin*. For episodes since 1950, the intensity is based on the pattern and magnitude of SST anomalies in the tropical Pacific. Since both warm and cold episodes tend to reach their peak during the Northern Hemisphere winter, they tend to span two different calendar years. In some cases years are listed as both cold and warm episode years (e.g., 1925, 1939, and 1976). In those cases there was a transition between the two extreme states during the course of the year. For more details concerning the intensity, onset and demise dates of the episodes since 1950, the reader is referred to the Climate Prediction Center's website. See (http://nic.fb4.noaa.gov/products/analysis_monitoring/ensostuff/ensoyears.html).

an El Niño episode evolves, significant changes occur in both the subsurface temperatures and the depth of the oceanic thermocline. In the early stages of El Niño episodes, the oceanic thermocline is deeper than normal in the western and central equatorial Pacific, in association with an abnormally deep pool of warm ocean water in those regions.

As El Niño episodes progress to the mature phase, the depth of the thermocline gradually decreases in the central and western equatorial Pacific and increases in the eastern equatorial Pacific in response to a reduced strength of the low-level easterly winds. As a result, subsurface temperatures become cooler than normal in the western equatorial Pacific and warmer than normal across the eastern equatorial Pacific. In the latter stages of El Niño episodes, as the heat content in the upper ocean is gradually depleted, both the depth of the thermocline and subsurface temperatures become less than normal throughout most of the equatorial Pacific. Accompanying this evolution, the warmer-than-normal temperatures in the eastern equatorial Pacific become increasingly confined to a shallow layer near the ocean surface, which sets the stage for a transition to either a neutral state or a La Niña episode. This transition process is critically dependent on the evolution of the low-level atmospheric winds. If the low-level easterly winds strengthen, the resulting increase in oceanic upwelling over the eastern equatorial Pacific brings cold ocean waters to the surface, as illustrated by blowing on the water in Figure 2-6. If this drop in SSTs is sufficiently large, it can lead to the onset of La Niña conditions.

Conversely, in the early stages of La Niña episodes, the thermocline is generally shallower than normal across the equatorial Pacific. The thermocline gradually deepens in the western Pacific during the mature phase of La Niña episodes, and in the central Pacific during latter stages of the episode. As a result, the subsurface temperatures become warmer than normal in these regions, while the ocean surface temperatures remain colder than normal. This decrease in the overall volume of abnormally cold ocean waters indicates an increase in the upper ocean heat content and results in conditions more favorable for a transition to either a neutral state or an El Niño episode. Once again the critical factors in the transition are the low-level winds and the subsurface temperature structure.

THE EVOLUTION OF EL NIÑO 1997–1998

Weak cold (La Niña) episode conditions prevailed during most of 1995 and all of 1996. During that period, enhanced low-level easterly winds contributed to a gradual build-up of warm water and upper ocean heat content in the western equatorial Pacific, which expanded into the central and eastern equatorial Pacific during January–March 1997 (Figure 2-7, top panel). This evolution is generally viewed as a precursor to El Niño. During this precursor period, cold episode conditions faded rapidly, as the sign of sea surface temperature (SST) anomalies switched from negative to positive in many sections of the equatorial Pacific (Figure 2-7, middle panel). At the same time, the equatorial low-level easterly winds became weaker than normal (Figure 2-7, 850-hPa u). These conditions prompted the Climate Prediction Center (CPC) to issue an ENSO Advisory on April 9, 1997 (solid time line in Figure 2-7), which stated that this evolution was consistent "with the possible onset of warm episode conditions."

Warm (El Niño) episode conditions continued to develop during April 1997, as the low-level easterly winds remained weaker than normal and the SSTs in-

"Howard! That's enough El Niño."

FIGURE 2-6. This cartoon, from a Sunday news magazine published in October 1997, reflects the nation's interest in the physical aspects of El Niño, which depends on winds blowing over the warm ocean waters to transfer energy and moisture into the atmosphere. (Copyright 1998 Mike Twohy, *USA Weekend*; reprinted with permission)

creased to near 28°C across the eastern and central equatorial Pacific. The CPC issued another ENSO Advisory on May 9 (dotted time line in Figure 2-7), stating that this evolution "indicates the early stages of a warm episode." SSTs continued to increase during May, which is a period when SSTs normally decrease throughout the central and eastern tropical Pacific. As a result, SST anomalies increased rapidly throughout this region, especially near the South American coast where the departures exceeded +4°C by early June.

Consistent with this warming, convection and rainfall (Figure 2-7, OLR values less than or equal to 240 W m^{-2}) developed over the central and eastern tropical Pacific. At the same time, convective activity weakened and drier-than-normal conditions developed over Indonesia and sections of the eastern Indian Ocean. Also, the low-level equatorial easterlies weakened dramatically, and by early June westerlies (west-to-east flow) appeared over much of the equatorial Pacific. This evolution was accompanied by a continued eastward expansion of the western Pacific warm pool, and by early June SSTs were greater than 29°C everywhere from Indonesia east to near 140 W.

These developments prompted the CPC to issue ENSO Advisories on June 10 (dot-dash time line in Figure 2-7) and on June 26, indicating that the El Niño

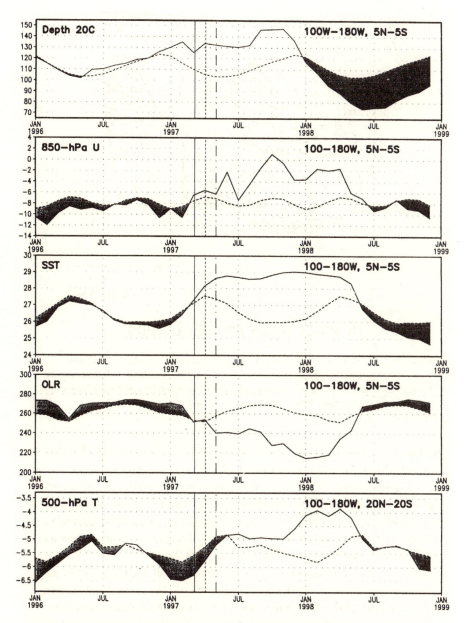

FIGURE 2-7.   Time series of the depth of the 20 C isotherm in the ocean (a measure of the thermocline depth), the low-level (850-hPa) east-west component of the wind, the sea surface temperature, the outgoing longwave radiation (OLR, a measure of convective activity and rainfall), and mid-tropospheric (500-hPa) temperature during 1996–1998. The regions over which the variables are averaged are indicated in the upper right of each panel.

conditions were intensifying and that further intensification was likely. The June 26 Advisory stated that "strong warm episode conditions have developed" and that the evolution during early 1997 was similar to that observed in previous strong warm episodes. This Advisory also stated that warm episode conditions would likely continue through the remainder of 1997. On the basis of these predictions, the CPC held a series of press conferences and briefings to inform emergency and water resource managers as to the expected impacts over the United States.

SST anomalies also increased in the subtropics and in the lower mid-latitudes of the eastern Pacific in both hemispheres from April to June. During this period, persistent weaker-than-normal high pressure systems to the west of Chile and California, features characteristic of warm episodes, resulted in weaker-than-normal coastal upwelling and the anomalous advection of tropical water into these regions.

SSTs reached record levels throughout the central and eastern equatorial Pacific during August–December 1997, as departures exceeded +1°C from the date line east to the South American coast, +4°C east of 140° W, and +5°C near the Galapagos Islands. The SSTs then remained nearly constant through early May 1998, despite a decrease in anomalies during early 1998. This decrease in the anomalies did not represent a weakening of warm episode conditions, but rather reflected nothing more than the normal increase in the climatological mean temperatures across the eastern half of the equatorial Pacific that occurs during this period.

Consistent with the extremely warm SSTs, enhanced convection and rainfall persisted over the entire tropical eastern Pacific from May 1997 through May 1998 (Figure 2-7, OLR panel). During this period, suppressed convection and drier-than-normal conditions dominated Indonesia, Malaysia, and the eastern Indian Ocean (Figure 2-8). A weaker pattern of drier-than-normal conditions was also observed over Central America and northern South America. These very warm SSTs, in combination with the intense convection, acted to heat the overlying tropical troposphere (the lowest 12 km of the atmosphere). As a result, mid-tropospheric tropical temperatures were higher than normal from July 1997 through June 1998 (Figure 2-7, bottom panel).

Beginning late in 1997, the depth of the thermocline began to decrease, and it actually became shallower than normal by early 1998 (Figure 2-7). During this period, the abnormally warm ocean temperatures became increasingly confined to near the ocean surface. Below this layer, ocean temperatures became increasingly colder than normal. This depletion of the upper ocean heat content signified an eventual end to the El Niño episode. However, as mentioned earlier, the timing of the transition phase-out of El Niño episodes depends not only on the subsurface thermal structure but also on the strength of the low-level easterly winds. These low-level winds remained very weak through April. As a result, the SSTs remained high, and atmospheric circulation features continued to reflect strong El Niño conditions.

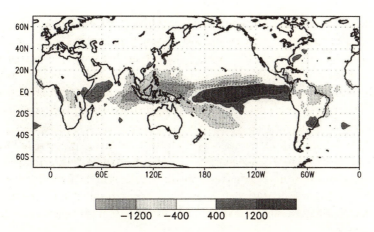

FIGURE 2-8. Precipitation departures from normal (mm) for the period May 1997–May 1998. Data are obtained by merging raingauge observations and satellite-derived precipitation estimates (Xie and Arkin 1997). Departures are based on the 1979–1995 base period means. Contour interval is 400 mm.

Early in May, the strength of the low-level easterly winds increased, which induced upwelling that quickly brought colder subsurface waters to the surface. This cooling, combined with the warmer waters farther west, produced conditions favorable for the maintenance and further strengthening of low-level easterlies in this region. It also led to a weakening of convection over the central and eastern equatorial Pacific and a strengthening of convection over Indonesia. Thus, a feedback process was established whereby the increase in strength of the easterlies led to cooling of the ocean surface and a weakening of convection over the region. This then led to increased convection over Indonesia, which favored an intensification of the easterlies over the central equatorial Pacific and a cooling of the ocean to the west of where it was first observed. This process occurred rapidly during May and June 1998, resulting in a rapid transition from El Niño to La Niña conditions.

WEATHER RESULTING FROM EL NIÑO
1997–1998 OVER NORTH AMERICA

Over North America, typical warm (El Niño) episode impacts on weather conditions began to appear late in 1997. Wetter-than-normal conditions developed along the southern tier of the United States during November and December 1997 and continued in many sections until April 1998 (Figure 2-9). During this period, rainfall amounts in many sections of California and the Gulf Coast were 200 percent to 300 percent above normal. This heavy precipitation was associated with

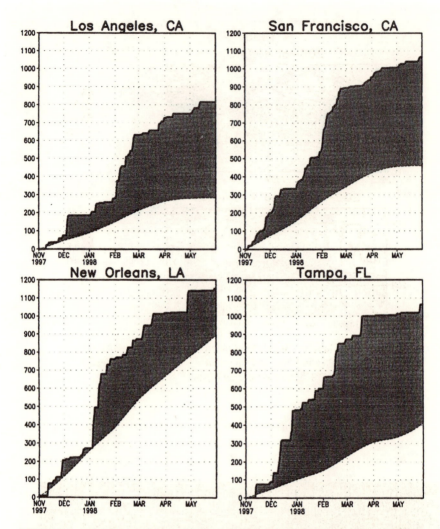

FIGURE 2-9. Observed and normal accumulated precipitation (mm) for selected stations in California and along the Gulf Coast. The accumulated observed precipitation is indicated by the heavy solid line. The accumulated normal precipitation is indicated by the thin dashed line.

a stronger-than-normal wintertime jet stream that extended east across northern Mexico and the northern Gulf of Mexico (Figure 2-10, top). This jet stream was also an important factor leading to increased storminess and several severe weather outbreaks in sections of the Southeast from January through March 1998, as illustrated in Figure 2-11.

The overall southward shift from the normal position of the jet stream and the storm track over North America contributed to less storminess and precipi-

JFM 1998
200−hPa Wind

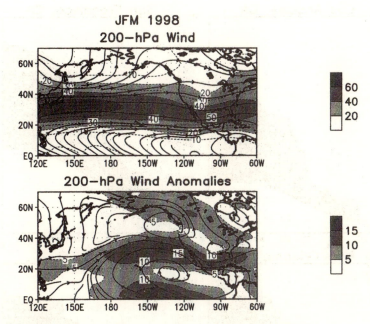

200−hPa Wind Anomalies

FIGURE 2-10.   January–March 1998 mean (top) and anomalous (bottom) 200-hPa wind flow and speed (m s⁻¹). Anomalies are departures from the 1979–1995 base period means.

FIGURE 2-11.   One of the strong tornadoes occurring during an outbreak of tornadoes across Arkansas and Tennessee on April 16, 1998. These were attributed to El Niño's influence on atmospheric conditions. (Courtesy Charles Bright)

tation over the extreme northern United States and southern Canada, as well as to significantly fewer cold-air outbreaks across the region. Also, with the main storm track being so far south, storms were detached from the very cold polar air that remained locked up in the Arctic. As a result, southern Canada and the United States were dominated by relatively mild air masses that moved from the eastern Pacific east over the continent (Figure 2-12).

These warmer-than-normal conditions developed late in 1997 over the northern Plains and the upper Midwest of the United States and over western and

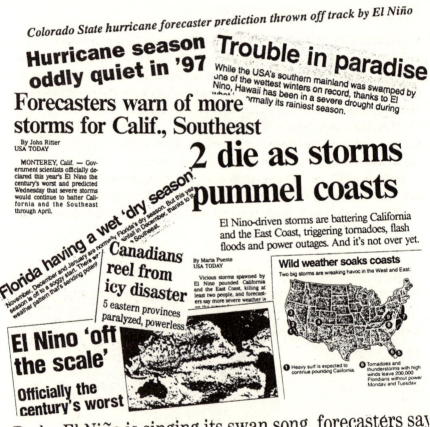

FIGURE 2-12. An assembly of newspaper headlines focusing on the weather events and climate conditions caused by El Niño, including various forecast outcomes.

central Canada. Although there was considerable month-to-month variability in the pattern, temperatures during the period from December 1997 through April 1998 were on average much warmer than normal over the northern half of the United States and most of Canada (Figure 2-13). This abnormal warmth, together with the lack of storminess, resulted in less-than-normal snowfall. Figure 2-14 shows areas that received snowfall totals 4 inches (10.2 cm) or more below normal; snowfall was well below normal over large portions of the northern United States, including the northwest, the upper Rocky Mountains, the Midwest, and the Northeast.

SEASONAL CLIMATE OUTLOOKS

Seasonal climate outlooks at the Climate Prediction Center are made in the middle of each month. These outlooks are based on statistical and numerical guidance, with an ever-growing emphasis on physically based methods. In particular, forecasts of the departures from normal of equatorial Pacific sea surface temperatures now play an important part in the forecast process, and the ENSO cycle is a major source of forecast skill. Also the ENSO-related circulation features, such as changes in the wintertime jet stream and in North Pacific blocking activity, are major physical aspects of these forecasts.

By June 1997, the NCEP prediction models (both physical and statistical) indicated that strong El Niño conditions were likely to continue in the tropical Pacific into early 1998. Previous studies and ongoing research indicated that

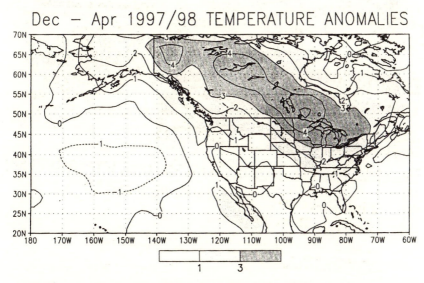

FIGURE 2-13. Temperature anomalies (departures from normal) for the period December 1997–April 1998. Units are degrees C. Contour interval is 1°C.

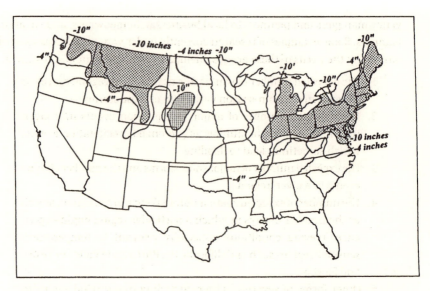

FIGURE 2-14.   Areas of the nation receiving winter 1997–1998 snowfall amounts that were 4 inches or more (10.2 cm) below normal. Much of the northern half of the nation had much less snow than usual. Certain northern areas within the shaded zones (10 inches [25 cm] or more below normal) had 20 to 40 inches (50–100 cm) less than their winter normals. However, the mountains of California, Utah, and Colorado had much-above-normal snowfall with 600 inches (1,524 cm) recorded at some of the higher elevations in the Sierra Mountains of California.

the El Niño impacts on the weather of the United States would begin during September–November, would be strongest during the heart of the northern winter (January–March), and would then gradually diminish during April and May. Examples of the official forecasts issued are shown in Chapter 5.

Since the developing El Niño was already strong in June and expected to remain strong into early 1998, Climate Prediction Center forecasters began to base their predictions for the upcoming winter on composites of the weather impacts observed during previous strong El Niño episodes. The resulting skill for the temperature outlooks for the winter (December–February) season was near or above 50 percent for all forecasts made from June through November. In addition, the skill of the winter precipitation outlooks ranged from 30 to 50 percent for all forecasts made during the same period. This high level of skill for both the long-range temperature and the precipitation forecasts was unprecedented.

LESSONS LEARNED

On the basis of our experience with El Niño 97–98, we now recognize the need for more detailed long-range predictions that are tailored to specific user groups ranging from government decision makers, water resource managers, and emergency managers and planners to local businesses, homeowners, and vacation planners. Some examples of research currently under way, which was spurred directly by the needs of these groups, include the development of statistics on:

1. The expected number of winter storms in a given region, as well as the expected within-season variability of these storms
2. The expected number of major precipitation events in a given region, as well as the number of prolonged precipitation events, which sometimes lead to flooding
3. The expected number of major snow storms and the number of snow events in a given region
4. The number of expected cold-air outbreaks in a given region, as well as the number of days in which a particular region might experience freezing conditions, especially relevant to temperature-sensitive regions such as California, the Rio Grande Valley in Texas, and Florida
5. Other forms of severe weather, including major wind events in coastal regions.

In addition, there is a need to better understand the physical processes by which the ENSO cycle ultimately impacts the development and frequency distribution of eastern Pacific and Atlantic basin tropical storm and hurricane activity.

Since there is considerable event-to-event variability in the strength of ENSO, it is necessary to better understand how this variability affects the structural characteristics of the atmospheric circulation pattern, which ultimately affects the pattern and the intensity of related impacts on temperature and precipitation. Improved long-range dynamical prediction of these conditions is contingent upon realistic and accurate simulations of the ENSO-related atmospheric circulation features, such as the changes in the subtropical ridges and their impacts on the wintertime jet streams. A correct dynamical simulation of these features ultimately involves an accurate simulation of the tropical rainfall and heating patterns and the interaction between these conditions and the extratropical atmospheric dynamics.

REFERENCES

Aceituno, P. 1988. On the functioning of the Southern Oscillation in the South American sector. *Mon. Wea. Rev.*, 116, 505–525.

Arkin, P. A. 1982. The relationship between interannual variability in the 200 mb tropical wind field and the Southern Oscillation. *Mon. Wea. Rev.*, 110, 1393–1404.

Berlage, H. P. 1966. The Southern Oscillation and world weather. *K. Ned. Meteorol. Inst., Meded. Verh.*, 88, 152pp.

Bhalme, H. N., Mooley, D. A., and Jadhav, S. K. 1983. Fluctuations in the drought/flood area over India and relationships with the Southern Oscillation. *Mon. Wea. Rev.*, 111, 86–94.

Bjerknes, J. 1966. A possible response of the atmospheric Hadley circulation to equatorial anomalies of ocean temperature. *Tellus,* 18, 820–829.

———. 1969. Atmospheric teleconnections from the equatorial Pacific. *Mon. Wea. Rev.,* 97, 163–172.

Caviedes, C. N. 1973. Secas and El Niño: Two simultaneous climatological hazards in South America. *Proc. Assoc. Amer. Geogr.,* 5, 44–49.

Gadgill, S., Joseph, P. V., and Noshi, N. V. 1984. Ocean-atmosphere coupling over the monsoon regions. *Nature,* 312, 141–143.

Halpert, M. S., and Ropelewski, C. F. 1992. Surface temperature patterns associated with the Southern Oscillation. *J. Climate,* 5, 577–593.

Hastenrath, S., and Heller, L. 1977. Dynamics of climate hazards in northeast Brazil. *Quart. J. Roy. Meteor. Soc.,* 103, 77–92.

Kiladis, G. N., and Diaz, H. F. 1989. Global climatic anomalies associated with extremes in the Southern Oscillation. *J. Climate,* 2, 1069–1090.

Kousky, V. E., Kagano, M. T., and Cavalcanti, I. F. A. 1984. The Southern Oscillation: Oceanic-atmospheric circulation changes and related rainfall anomalies. *Tellus,* 36A, 490–504.

Mossman, R. C. 1924. Indian monsoon rainfall in relation to South America weather, 1875–1914. *Mem. Indian Meteor. Dept.,* 23, 157–242.

Rasmusson, E. M., and Carpenter, T. H. 1983. The relationship between eastern equatorial Pacific sea surface temperatures and rainfall over India and Sri Lanka. *Mon. Wea. Rev.* 111, 517–528.

Ropelewski, C. F., and Halpert, M. S. 1986. North American precipitation and temperature patterns associated with the El Niño-Southern Oscillation (ENSO). *Mon. Wea. Rev.,* 114, 2352–2362.

Ropelewski, C. F., and Halpert, M. S. 1987. Global and regional scale precipitation patterns associated with the El Niño/Southern Oscillation. *Mon. Wea. Rev.,* 115, 1606–1626.

———. 1996. Quantifying Southern Oscillation-precipitation relationships. *J. Climate,* 5, 1043–1059.

Troup, A. J. 1965. The Southern Oscillation. *Quart. J. Roy. Meteor. Soc.,* 91, 490–506.

Walker, G. T. 1923. Correlation in seasonal variations of weather. Part 8: A preliminary study of world weather. *Mem. Indian Meteor. Dept.,* 24, 75–131.

———. 1924. Correlation in seasonal variations of weather. Part 9: A further study of world weather. *Mem. Indian Meteor. Dept.,* 24, 275–332.

———. 1928a. World Weather III. *Mem. Roy. Meteor. Soc.,* 2, 97–106.

———. 1928b. Ceará (Brazil) famines and the general air movement. *Beitr. Phys. d. freien Atmos.,* 14, 88–93.

Walker, G. T., and Bliss, E. W. 1932. World Weather V. *Mem. Roy. Meteor. Soc.,* 4, 53–84.

Wyrtki, K. 1975. El Niño—the dynamic response of the equatorial Pacific Ocean to atmospheric forcing. *J. Phys. Oceanogr.,* 5, 572–584.

Xie, P., and Arkin, P. A. 1997. Global precipitation: A 17-year monthly analysis based on gauge observations, satellite estimates and numerical model outputs. *Bull. Amer. Meteor. Soc.,* 78, 2539–2558.

3

 ## Was El Niño A Weather Metaphor—A Signal For Global Warming?

LEE WILKINS

his chapter reviews media coverage of El Niño 97–98 and identifies some significant trends within that coverage. The coverage analyzed includes that provided by the major American television networks, the elite press, and significant regional newspapers. During the early months of the study period, news coverage of El Niño was focused on the science of the prediction and was framed as an issue of risk with appropriate uncertainty. However, as the predictions themselves were borne out in real-world phenomena, coverage of El Niño became event driven, and the phenomenon itself was treated as certainty. The risks of climate change attributed to El Niño outweighed the potential benefits in many media reports. Coverage of El Niño was extensive, particularly on the West Coast of the United States, where many individual weather events were connected with the larger phenomenon. The chapter then explores the possibility that the totality of the media coverage may have two lasting impacts. First, on the basis of existing scholarship on mass communication and risk communication, it is reasonable to suggest that the extensive news coverage of El Niño may have had some influence on public perception of climate change, particularly the salience of climate change in discrete regions of the nation. Second, the chapter suggests that the mediated reality of the 1997–1998 event will serve as a signal event for popular and political understanding of the consequences of global warming.

WHO WROTE THE EL NIÑO NEWS?

Historically, journalism has been both hampered and helped by its definition of news. Previous studies of media coverage of a variety of "risky" events have noted that news accounts tend to be event focused, lack context, and treat science as a matter of dueling opinions, rather than a process of knowledge acquisition. These scholarly findings, which are long-standing, have had some impact on the professional community, particularly among science writers, who over the past two decades have become both better trained in science and more aware of the limitations of the concept of "news"—at least when it comes to reporting certain sorts of events.

Media coverage of El Niño, in general, reflected these previously documented trends. This study was based on an extensive content analysis of news coverage of El Niño 97–98 found in a diverse group of American newspapers and on American network televison. The content analysis of news papers included the *Los Angeles Times*, the *Seattle Times*, the *New York Times*, the *Washington Post*, and the *Chicago Tribune*. The televison programming examined included the three major networks, the American Broadcasting Company (ABC) with fifty-nine programs analyzed, the National Broadcasting Company (NBC) with fifty-eight programs analyzed, and the Columbia Broadcasting System (CBS) with seventy-two programs analyzed. The number of news stories studied for their contents is shown in Table 3-1, which shows that 1,595 news stories were analyzed, as were 189 broadcast stories.

News coverage begins with individual journalists, most of whom function anonymously. It is important to note that the sort of reporters who covered the El Niño story changed over time. This change was borne out by both the content analysis and an informal survey of science writers who gathered in February 1998 in Philadelphia at the annual professional meeting for science journalists. The science writers were asked how much they had written about El Niño and, when appropriate, when that coverage had appeared. In the majority of

TABLE 3-1.   Number of articles about El Niño analyzed for content in five national newspapers from June 1, 1997, through June 1998.[a]

Month	Los Angeles Times	Seattle Times	Chicago Tribune	Washington Post	New York Times
June 1997	4	2	0	1	4
July	12	2	1	0	7
August	24	4	2	3	9
September	55	10	8	1	13
October	70	37	10	4	21
November	120	25	11	9	27
December	138	20	12	11	31
January 1998	90	25	19	2	41
February	232	36	37	10	43
March	109	29	36	5	24
April	68	7	9	2	11
May	42	1	3	2	4
Totals	964	198	148	50	235

[a]Nonnews stories were not included.

cases, science writers said they had written relatively few El Niño stories and that those they had written had focused primarily on the "science" of understanding the phenomenon and on the probabilities and weather consequences of its existence.

The content analysis of nearly 1,800 news articles confirmed this view. Early coverage of El Niño, the converge that focused on the science of the phenomenon and framed climate predictions in more problematic terms, was written generally by science and environmental writers. As shown by data in chapter 4, these early articles usually cited scientists as sources for some or all of the information presented. These stories appeared primarily in prestige publications such as the *Los Angeles Times* and the *New York Times*. The *New York Times*, which published much less in total about El Niño than did the *Los Angeles Times*, tended to place these early stories on its science pages. The *Los Angeles Times*, which produced more news coverage of El Niño than any other publication analyzed, placed these early stories throughout the entire paper. But, aside from the differences in placement—which certainly reflects the Los Angeles's paper's greater editorial commitment to environmental coverage—these early stories in both publications shared many characteristics. They are models of quality science journalism, cautious about the inevitability of any scientific prediction, thorough in the citation of scientific sources with differing views, and not the least alarmist or causal in tone. The headlines in Figure 3-1 help illustrate the uncertainties tied to the El Niño-related seasonal climate forecasts.

However, by late fall 1997, El Niño stories were seldom being written by science and environmental writers. Some of the change reflects the real, event-oriented world in which journalists work. California mud slides, for example, were covered by reporters who wrote about individual communities or by reporters assigned to other areas and who were pulled off their normal assignments to write about quick-onset events. The same was true for the *Denver Post* in its coverage of the October 1997 blizzard in the High Plains; only a minority of the ensuing stories were written by science and environmental journalists. As the winter, and the weather impacts attributed to El Niño, mounted, proportionately fewer stories about El Niño were written by science writers, regardless of media organization. Network television, which employs relatively few science journalists and which tended to cover El Niño almost exclusively as a series of discrete, disastrous events, also followed this pattern.

These later stories also demonstrate a change in focus. While the earlier, science-based stories tended to center on El Niño itself, in most later news coverage El Niño received only a glancing mention. Instead, El Niño became the *cause* of other events that took center stage in individual news accounts. Part of this shift reflects journalistic craft attitudes in defining news. If a house is sliding down a hill in southern California, or a California beach is littered with storm-caused debris (Figure 3-2), that event is the focus of the news story—not the storms or the heavy rains that were the root of the problem in the preceding days.

Weather uncertainties loom

WARM NORTH ATLANTIC WATERS MAY HELP MAKE THIS EL NINO EVEN SNOWIER IN THE EASTERN STATES - ONE OF THE BIG DIFFERENCES THIS YEAR FROM 1982/83

Hot year for 1998

Warm El Nino winter could mean early spring planting

We're hearing plenty of hype about the strong weather phenomenon known as El Nino that could bring flood, droughts and hurricanes around the world.

In Illinois, it's likely to bring less snowfall and milder winter temperatures, notes Steve Hollinger, with the Illinois State Water Survey. There is a very high probabil-

we should be in pretty good shape through March, says Hollinger.

"Illinois farmers may be able to get crops in a little earlier than normal in the spring," says Hollinger. Farmers who planted this year's crops in June would welcome an early start to the 1998 growing season. "Overall this can be rather positive for

Value of El Nino forecasts gets mixed reviews

■ Scientists seek refined effort

La Nina turmoil likely to continue through June

By Shannon Tangonan
USA TODAY

Blame La Nina. El Nino's de-

Weather effects o
La Nina, cooler-than-normal sea-surfac

Meteorologist sees weather woes ahead

STUDY BY NOAA SCIENTISTS COULD LEAD TO BETTER PREDICTIONS OF EL NIÑO, COMMERCE AGENCY SAYS

Experts hedge El Nino forecast

Special Report

Science struggling to predict El Niño devastation

WEATHER: Rainfall this winter could be three or four times the normal amount, or not.

Pesky El Niño is singing its swan song, forecasters say

The Fury Of El Nino

Suddenly nobody's calling it El No-Show anymore. What have we learned from the climate event of the century?

O nly a few months ago, El Nino was starting to look like the most overhyped story of the decade. The periodic warming of Pacific

TIME

FIGURE 3-1. The print media focused attention on the many forecasts of future weather and climate conditions issued from June 1997 to May 1998. This assortment reveals the variety of forecasts issued, the frequent differences and uncertainties expressed, and the questioning of forecasts that were issued.

FIGURE 3-2.   The print media presented numerous photographs of damage from El Niño–driven storms to illustrate the variety of problems it caused. This photograph shows a California beach at Monterey Bay covered with mountains of debris that had been washed ashore during a February 1998 coastal storm. (Courtesy Robert A. Eplett, California Office of Emergency Services)

This shifting of focus also supported another change in El Niño coverage. Early, science-based stories about El Niño, those that tended to be written by science writers, tended not to focus on the "bad consequences" of the phenomenon but rather discussed a series of possible consequences, each one seen as a possibility, not a certainty. However, when event-oriented coverage dominated, as it did from October 1997 through most of the study period (to June 1998), the consequences became certainties, and almost all of them (except the mild winter in the North) were bad. This was particularly true for the coverage of policy-oriented stories, almost all of which were written by non-science journalists and almost all of which assumed, at least in the context of individual news accounts, that El Niño was the "cause" of particular problems.

It would be inappropriate to "blame" news organizations for failing to have their science writers produce all El Niño-related coverage. Indeed, larger news organizations often asked their science writers to contribute paragraphs explaining El Niño for other news stories, at least according to the science writers surveyed early in 1998. But, in a journalistic sense, and in a profession with daily and sometimes hourly deadlines and a series of events that, in some parts of the nation, had serious consequences in terms of property damage, there were simply not enough science writers to complete the journalistic task. In addition, scholars have documented that many sorts of stories about risk tend to be framed in terms of bad consequences (Singer and Endreny, 1987). Thus, these elements of news coverage followed established journalistic patterns.

Thus, during the study period, the El Niño story progressed from one of scientific controversy to one of weather-related problems, including everything from minor inconveniences to individual catastrophes to state or region-wide problems that had to be addressed. No longer was El Niño framed as a scientific phenomenon; instead, it represented a collection of problems with unfavorable consequences from the individual through the system level.

Differences among News Organizations

Scholars have noted that individual news organizations tend to cover individual events on the basis of their sense of audience and mission within a community. This is particularly true of newspapers, including those with national circulation like the *Los Angeles Times* and the *New York Times*. And, coverage of El Niño reflected these findings.

In general terms, all the newspapers studied focused their coverage on specific events. El Niño coverage peaked in the months of November, December, January, and February in all publications studied and dropped off the news agenda early in the month of May 1998. This period of peak coverage coincided with the rainy season on the West Coast and with winter throughout the remainder of the country. Because of its focus on the environment, an issue of public interest in California, and because of the series of discrete events that occurred on the West Coast, the *Los Angeles Times* provided more than three times the coverage of the phenomenon of any other of the five papers analyzed. However, it is important to note that in more than 75 percent of these stories, El Niño was mentioned only once or twice, while the focus was on the sort of property damage that resulted from heavy rains or floods. As shown in Table 3-1, the least coverage of the phenomenon was provided by the *Washington Post*. The *Washington Post*, because it is located in an area that experienced relatively few bad weather-related consequences, provided the briefest coverage—some early discussion of the science and then stories that focused on what was happening "out west" and some stories that focused on possible policy responses. However, the policy-related stories generally focused on congressional or executive meetings, travel, or other specific events, rather than on issues of preparedness.

It is also important to note what the news stories did *not* cover, and again this followed a well-documented pattern. Scientists have often criticized journalists for omitting qualifying details in science stories. These details, which are essential to scientific understanding, generally do not fit in headlines and are considered somewhat superfluous to working definitions of news, particularly under deadline pressures. This same pattern was particularly prevalent when it came to news reports of warnings about El Niño–generated conditions. Chapter 4 documents the various academic institutions and the federal and state agencies that produced different reports with contents that were certainly distinctive in the scientific sense. But when those predictions reached the mass media, both their sources and some of the specific content became blurred. News accounts sourced the stories with phrases such as "the National Weather Service" or "local scientists," rather than draw fine distinctions between and among the various agencies actually involved in the prediction process. This blurring of forecast messages was true of most publications analyzed and even more true of national television, which usually sourced individual stories with phrases such as "scientists say" or "weather predictions" (see Figure 3-1). For the news organizations involved, the essential element was the warning message itself, rather than the agency that was releasing the information. Further, once El Niño became an accepted "fact" in the news, the distinctions were omitted almost entirely.

The most distinctive sort of coverage of El Niño was provided by the *Chicago Tribune*. But, that coverage also reflected that paper's mission. While the *Tribune* reported on various weather-related events, it also tied El Niño into economic consequences, particularly those in the nation's heartland where the weather has major impacts on energy use and on crop production and the nation's financial investment in farming. Thus, the *Tribune* printed a larger proportion of El Nino stories that focused on the economic consequences for consumers, farmers, and the commodities markets than did any of the other publications studied. Again, it is important to note that these stories tended to assume that El Niño was "fact" and was the cause of a variety of weather-related events.

The only other publication that placed a similar, but not as pronounced, emphasis on the economic consequences of the phenomenon was the *New York Times*. Several articles assessed the diverse variety of consequences occurring across the nation.

News accounts early in the study period tended to quote scientists more frequently than did news accounts beginning in November and continuing through June 1998. Again, this sourcing pattern reflects the kinds of stories that were being written, as well as journalistic routines. When covering heavy rains in Seattle, it is reasonable that journalists should quote public officials involved in hazard response. Similarly, stories that focused on commodities trading quoted sources in the financial sector rather than scientists. Stories that focused on policy making, particularly those printed in the *Washington Post*, tended to quote public officials who were more concerned with government response than with the science of the underlying events. The same was true for stories that appeared in the *Los Angeles Times*, which focused on state and regional responses to the various events in California. Thus, in a journalistic sense, individual forecasts and warning messages became detached from a variety of news events that constituted a response to El Niño. The individual publications studied did not differ from one another in this respect.

These sourcing patterns reflect two important journalistic norms: an event-driven news agenda and news values that promote local as opposed to national sources. The various newspapers studied sought local experts on a variety of fields, rather than relying on national scientific experts. Furthermore, as El Niño became an accepted "fact," there was less and less journalistic reason to seek information from the scientific community and a compelling need to write about El Niño's impact on local people, most of whom were not familiar with the science of the phenomenon but were instead more concerned about local activity.

Finally, it is important to note there was something of a geographic bias in the news coverage—a bias that reflects the fact that all national television newscasts originate on the East Coast and that the *New York Times*, a newspaper that did not distinguish itself in either the amount or the quality of its El Niño coverage, is viewed as sometimes setting the national news agenda for both print and broadcast reports. Put most simply, news coverage of El Niño was a West Coast phenomenon. Papers in the western part of the country—specifically the *Seattle Times* and the *Los Angeles Times*—printed significantly more El Niño stories (see Table 3-1) than most of the other publications studied. During November and December, when heavy rains caused a variety of problems on the West Coast, papers in the East devoted comparatively little space to El Niño. What stories did run tended to be wire service accounts. Stories from the Midwest tended to cluster somewhat later in the study period, coinciding, not surprisingly, with the mild winter and the resulting impacts, including lower heating costs and record retail and home sales (see chapter 6). Television coverage followed an event-centered pattern, with a focus on El Niño–caused problems in November, December, and January. Figure 3-3 pokes fun at TV weathermen who were the typical providers of news about bad El Niño weather. Most of these stories, however, originated on the West Coast, and all were event related.

Thus, in a national sense, El Niño never dominated the news agenda in the prestige East Coast publications in the way that it dominated the news coverage of those in the West. At one level, this geographic emphasis reflects where actual events were occurring in different parts of the country. But, at another level, a relative lack of coverage by the papers in the East also meant that policy elites, which tend to read those papers, saw less news about El Niño than might have been appropriate, considering the scientific and climatological issues El Niño raised.

THE STORIES WITHIN THE EL NIÑO STORY

While ultimately almost 2,000 stories were individually analyzed on a variety of criteria,[1] it is important to look at the chronology of that coverage as well as at the individual pieces. In fact, one of the most significant findings of this study was the sheer volume and diversity of news coverage that included a mention of El Niño. However, the totality of that coverage tells a story that individual reports do not.

So we see that the true cause of bad weather, contrary to what they have been claiming all these years, is TV weather forecasters, who have also single-handedly destroyed the ozone layer via overuse of hair spray.

FIGURE 3-3.   This cartoon, from a Sunday news magazine published in December 1997, points out the ubiquitousness of El Niño information, including the blame assigned to El Niño for weather problems, presented by the print and television media. (Copyright Tribune Media Services, Inc., all rights reserved; reprinted with permission)

Science and Uncertainty Came First

At the beginning of the study period (June 1997), which was shortly after the first prediction of the emerging El Niño was reported in the popular media, news articles about El Niño focused almost exclusively on the science of the predictions and the complexity of the processes. For example, articles routinely noted that the predictions were based on buoys placed in the Pacific Ocean to monitor ocean temperatures. Further, this monitoring network was new, as was the scientific understanding of the ocean-atmosphere interactions. Thus, this was the first time a genuine El Niño prediction based on scientific data and new knowledge had been issued. Newspaper articles, whether they appeared in the *New York Times* or the *Los Angeles Times*, treated the topic as a scientific issue, and the predictions themselves were seen as open to re-evaluation, as illustrated in Figure 3-1. Consequently, these new accounts carried comments from scientists who said they did not subscribe to the prediction or, more frequently, couched the forecast in terms of probabilities that something might happen (see section on scientific differences over the predictions, chapter 4).

The science of El Niño was relatively new, and, while the 1997–1998 event was the largest yet recorded, scientists were not certain about the potential impacts on atmospheric circulation and the weather around the globe (see chapter 4). In the parlance of scholarship, these news stories fit neatly into what scholars have documented about many sorts of risk communication framed in the news media:

- They were event driven
- They presented the prediction as something that was open to revision when additional data were collected
- They quoted scientists who noted the such predictions tend to be completely accurate.

The first real break in this pattern of the scientific presentations in the media came in October, when an early-season snowstorm hit the mountainous regions of the West and the High Plains. News coverage of that event, as typified by the *Denver Post*, directly linked, in a causal fashion, the early, heavy snows with El Niño (see the cartoon from the *Rocky Mountain News*, Figure 3-4). These news reports circulated throughout the region and even made the

FIGURE 3-4.   This cartoon, from a September issue of a Denver newspaper, reflects two messages: (a) the efforts of some scientists and federal agencies to tie the problems of the 1982–1983 El Niño to those predicted for 1997–1998, and (b) the variations in the public's expectations for impacts from the El Niño weather according to individual interests and beliefs. (Reprinted with permission of Ed Stein, courtesy of the *Rocky Mountain News*)

national news agenda in the form of brief mentions on two of the three major televison networks. However, the El Niño connection was much less the focus of the television coverage than it was of the early print reports.

It took less than a week for a scientific revision to set in. Within that time, the Denver paper was reporting that scientists were arguing that the snowfall could not be directly attributed to El Niño and that, in addition, normal October weather patterns in the mountainous West often include significant early snows—although not to the snow depth or areal extent (the storm went east into Minnesota and Wisconsin) that was reached in 1997. (See chapter 4 for a description of the numerous scientific debates over which weather events of 1997–1998 to attribute to El Niño, and chapter 7 presents details of the October blizzard controversy). If readers had taken these October news reports at face value, they would have understood that those scientists who had linked the snowstorm to El Niño had published an alarm that further reflection and additional evidence discounted.

Climatologists, of course, have noted for years that regional variations in climate conditions that may occur are some of the most difficult elements to predict, as a result of global climate change due to human-induced warming (assuming one accepts the theory). Within the realm of regional variations of weather and climate conditions, a cause-effect relationship of individual weather events to larger climatological shifts—particularly when given data for a single year—is impossible to sustain in a scientific sense. This level of scientific uncertainty permeates scholarly discussions of climate change; it has also been documented that not much of this material reaches the daily news reports. The Denver case illustrates the fact that the scientific uncertainty, rather than being injected into each individual story, emerged as reporters went from covering specific events to writing about the causes of these events. What is remarkable is that some of the Denver news media concluded that El Niño was not a causal factor for the storm.

It was the last time that such a conclusion would dominate news coverage in the larger chronology of the thirteen-month El Niño story.

Uncertainty Shifts to Cause without Qualification

By November, when severe storms were beginning to reach the West Coast and sometimes followed a storm track east across the southern United States, a different narrative emerged. No longer was the El Niño prediction scientifically contentious in news reports. Instead, the meta-narrative focused on the specific events, or the impacts thereof, that were often linked directly to weather patterns either created or exacerbated by El Niño. Thus, the heavy rains due to El Niño led to mud slides, property damage, coastal erosion, traffic snarls, and school closings all along the West Coast where these events persisted for months. Figure 3-5, a cartoon published in many newspapers across the nation, illustrates how the California problems were being widely presented. While climatologists would certainly assert that winter is the rainy season in the West and that it is extraordinarily difficult to link an extra inch of rain on any given day to El Niño, this sense of bounded uncertainty failed to the permeate news reports after October. Instead, El Niño and its myriad consequences were treated as *fact*.

And, because the consequences were widespread—impacts occurred at most points over much of the North American continent during 1997–1998—news accounts reflected the diverse impacts of El Niño, but the precise coverage varied with the source. Thus, the *Chicago Tribune* attributed the Midwest's mild winter—and its many beneficial consequences—to El Niño with few qualifications. The *Los Angeles Times*, beginning in November and continuing through the rest of the study period (ending in June 1998), attributed a host of problems to the El Niño weather conditions. Furthermore, these were "picture-friendly" events, something the national television networks, all with Los Angeles bureaus, did not ignore. Thus, the entire country was treated to repeated pictures of the impacts of mud slides, beach erosion, gale-force winds, and lots and lots of rain all along the West Coast in stories that clustered in November, December, January, and early February. Figure 3-2 is typical of the photographs of California damages. In terms of narrative structure, El Niño, in the space of four months, had permutated from scientific predictions worthy of scientific study and debate to *climatological fact*.

FIGURE 3-5. Cartoonists captured the essence of many aspects of El Niño, including the many damage stories coming out of California, where storms and heavy rains caused numerous mudslides. (Reprinted with permission of Dana Summers, *Orlando Sentinel*)

Furthermore, it was climatological fact with real human—and animal—impacts. A single televison story broadcast late in winter illustrates the emerging El Niño narrative. The setting is a California beach, and the protagonist is an ailing sea lion. These animals—which were wonderfully telegenic—were washing up on California beaches in conditions too weak to survive. They were starving because their normal diet had been killed by the too warm waters of El Niño. The result was animal rescue teams with the unenviable job of telling well-meaning humans that the animals should not be "rescued" and that nature's cycle needed to continue without human interference. The story was broadcast as follows: Envision a close-up of big, brown baby sea lion eyes. Cut to environmentalists who were being interviewed, noting that these obviously distressed animals represented a "normal" environmental impact on wild populations. Then comes the caution from the on-camera journalist that well-meaning humans should not attempt to feed or otherwise rescue the dying animals . . . end of story.

The impact of this narrative is that El Niño is creating consequences that humans cannot alter. In this particular report, there was no discussion of any uncertainty about the causal relationship between El Niño and the sea lion population nor about the scientific debate surrounding such inferences. The relationship was a fact.

The focus on this individual broadcast news account is not to assert that journalists did not do their jobs or that they did their jobs badly. Television is a visually driven medium, and the seas lions were an appropriate focus for a news story that viewers—and, later, readers—would be interested in and respond to (note the cartoon in Figure 3-6). Rather, it is to suggest that, in the meta-narrative of news coverage of El Niño, scientific uncertainty regarding individual events was replaced with assertions of cause and effect. And, while assertions of the effects on individual events remained scientifically problematic (see chapter 4), in the larger scope of the science of global climate change, the accretions of diverse consequences clearly fit what most climate models have suggested might occur. It is an instance where, despite individual problems and differences, the totality of coverage reflected much of the current "scientific thinking" about the effects of global warming.

News coverage from early in the study period attempted to disengage individual events from the larger phenomenon. But, within months, these efforts were overwhelmed by the event—the center focus of individual news accounts. Almost all of them accepted El Niño and its climatological consequences as fact.

The sea lion story also illustrates another mediated phenomenon: the penetration of El Niño into the popular culture. Popular culture, unlike the more scientifically oriented work, reflects to society those issues that society has become, in some way, concerned with. Because popular culture arises in fictional as well as factual communications, there is no professional or craft demand for scientific accuracy. What is popularly accepted as fact, in a popular culture context, *is fact*, regardless of the "science" behind the issue.

FIGURE 3-6. The national news carried many stories about the excessive flooding caused by El Niño storms and heavy rains in California. This February 1998 cartoon aptly illustrates the problems and reflects the wide attention given to the problems with the seals and sea lions. (Copyright 1998 Washington Post Writers Group; reprinted with permission).

This emergence of El Niño into the popular culture had at least one tangible impact on the content analysis: humor and sports were eschewed. While this division reflects the traditional boundaries between news and other sorts of mass-mediated communication, the separation also results in an under-reporting of how many times El Niño worked its way into the mass media and into popular sorts of communication. Jay Leno did jokes about El Niño—at least until Monica Lewinsky became the focus of the national news (see Chapter 8). El Niño references found their way into the "Budweiser lizard" ad campaign. And, for several months, they ruled the sports section of the newspapers studied. For example, the Denver Broncos' chances of winning the Super Bowl in January 1998 were compared to the consequences of El Niño. Baseball and basketball stories suffered from an overload of El Niño metaphors.

While a non-news risk communication is beyond the specific focus of this study, it is very much at the core of the meta-narrative that emerged (Signoriellei and Morgan, 1990). In a popular sense, El Niño became the "cause" of any unexpected or unbelievable event, as illustrated in Figure 1-6. Furthermore, these varied weather outcomes and events gave a clear indication of the widespread impacts of the El Niño phenomenon. If popular culture is taken as one measure, albeit an imperfect one, of human concern with a particular issue, then El Niño had clearly captured the public's imagination. This public salience is crucial if policy makers attempt to make changes that demand public understanding and acceptance. Furthermore, scholars have documented that individuals tend to understand news accounts as part of a narrative frame and that this is particularly true of environmental news stories (Shanahan, 1993).

Thus, the original news story image of El Niño was transformed into something more symbolically charged. Contentious scientific prediction and uncertainties, as illustrated in Figure 3-1, became believable warnings, which were in turn transformed into a single cause for multiple sorts of effects. There was little discussion of the scientific basis of the phenomenon; instead, El Niño became a popularly accepted climatological fact. And, further, it was a "fact" that was accepted as having a variety of weather-related and climate-related consequences affecting the environment, society, and the economy. This narrative then moved into the popular culture. In the ecological system of mass-mediated messages, El Niño had moved from prediction to fact. It was the cause of diverse events. It sent a signal throughout society that something important—and far reaching—was changing.

WILL EL NIÑO 1997–1998 BECOME A SIGNAL EVENT?

A century ago, President Theodore Roosevelt defined our nation's great central task as "leaving this land even a better land for our descendants than it is for us." In his 1999 State of the Union Address, President Bill Clinton stated, "Our most fateful new challenge is the threat of global warming; 1998 was the warmest year ever recorded. Last year's heat waves, ice storms, and floods are but a hint of what future generations may endure if we don't act now. So tonight, I propose a new clean air fund to help communities reduce pollution, and tax incentives and investments to spur clean energy technology. And I will work with Congress to reward companies that take early voluntary actions to reduce greenhouse gases."

How this perspective on global warming ties to the El Niño event of 1997–1998 is a hypothesis now to be considered.

Journalists are people of both words and images. As such, they would be among those most able to accept the notion that the individual stories they produce are capable, over time, of forming a larger whole impression (Carey, 1989).

Mass communication scholars have been less ready to accept such an interpretation. Many studies, for example, suggest that people readily forget the facts of individual new stories, that opinion change on the basis of news coverage is fleeting, at best, and that people attend to stories they are personally interested in, making the formation of a public dialogue on emerging issues very problematic (Lowery and De Fleur, 1986).

However, other scholars have taken a more gestalt approach to evaluating how people understand the news. They have found, for instance, that readers and viewers approach news stories with psychological schemas and interpretation already in place (Graber, 1988). Further, these investigators claim that repeated narratives, particularly those in entertainment programming, can have an impact on vulnerable populations (Lowery and De Fleur, 1986) and that how news stories are put togther—the points that they emphasize and the points they omit—can influence individual interpretation of events (Iynegar, 1994). The totality of these findings argues that it is not the individual message so much as the accumulation of individual messages over time that has the capacity to influence thought and, sometimes, human behavior.

The literature of risk communications has traveled between these two diverse interpretations. Empirical research investigating individual responses to risk communication mirrors what is called the "minimal effects approach" to risk communication. In other words, the mass-mediated messages have relatively little impact on public perception of and public behavior concerning various risky events.

Beginning in the late 1980s, and continuing to the present, other scholars have argued for a more powerful effects approach to risk communication. This approach focuses on the long-term impact of multiple messages and analyzes the content of individual messages as important players in a multichannel, multilevel risk communication process involving not just audience members but political decision makers as well. This second approach to risk communication takes as its intellectual framework the politics of doing something about risk on both the individual and the system levels, instead of the earlier, more linear, behaviorist approach. It is from this more recent cohort of risk communication research that the concept of a signal event has emerged.

Phrased most simply, a "signal event" is a symbolic interpretation of an actual instance of risk that takes on a meaning that has both cultural and scientific overtones (Anderson, 1997). Signal events, or at least a rendering of them in the form of mass communication, become symbolically charged. They become, in a cultural and political sense, larger than the original occurrence and its direct societal and environmental effects.

Two of the most recent examples of international signal events are the Three Mile Island nuclear power plant accident and the Bhopal, India, chemical spill. The Three Mile Island disaster resulted in no loss of life and relatively little inconvenience to electrical power customers. Yet, Three Mile Island, in terms of the public evaluation of nuclear power, is credited with setting the stage for the decline of the nuclear power industry in the United States. In the sense that cultures construct meaning independent of events, the Three Mile Island event set the stage for American interpretations of the Chernobyl nuclear disaster. On the other hand, the Bhopal spill, which cost tens of thousands of lives and literally millions in various currencies around the globe (Wilkins, 1987), sent a signal to the American political establishment. Bhopal, scholars agree, resulted in profound political changes in terms of how the chemical industry was to be regulated in the late twentieth century. Right-to-know laws, the development of state hazardous materials legislation, and legislation requiring chemical manufacturers to make public the toxicity of the various compounds they produce are all a direct result of the "signal" that Bhopal sent to the American political system.

Of course, determining whether something has become a signal is not easy, particularly in the midst of an event. In fact, most signal events are recognized only in post hoc analysis when time and history have been able to add another layer of weight to symbolic meaning. Think, for example, of the changing cultural interpretations of AIDS or the rapidity with which some disasters, for example most tornadoes, are forgotten, while other disasters, such as the effects of the drug thalidomide, take on literally a life of their own.

Nonetheless, it is the contention here that the media coverage of El Niño 97–98 represents a potential cultural and political signal, but not of the warming of the oceans or the potential for better climate predictions. Instead, El Niño 97–98 may become the signal for the warming of the globe. Chapter 4 explores the El Niño-global warming relationship from a scientific basis and describes the scientific controversy that evolved during the event.

President Clinton, in his 1999 State of the Union address, and Vice President Gore, who claimed at an October 1997 conference that El Niño 97–98 was a result of global warming, may have been loose with the scientific facts about the ENSO phenomenon and global warming. But, as has been the case for the much of his presidency, Clinton showed an extraordinarily acute understanding of political symbolism. While individual Americans as voters, or their representatives in the public establishment, may have a difficult time conceiving of the fifty- or one-hundred-year probabilities of global climate change, a year of weird weather is much easier to grasp and provides more tangible instances for individual and system-level action. This cultural and political signal is not the result of any one news story, but rather of what scholar James W. Carey (1989) has called the ecology—or totality—of news coverage itself. Understanding the meta-narrative is one key to understanding why the news coverage of El Niño may symbolize much more than what happened between June 1997 and June 1998.

CONCLUSION: THE IMPACT OF A BELIEVABLE PREDICTION

At the level of the individual news account, coverage of El Niño followed patterns that have already been thoroughly described in risk communication and science communication literature. Taken at the individual level, journalists behaved in ways that were predictable and that have been well documented. But, at the system level, the El Niño coverage raises some important issues for both scientists and policy makers: namely, what is the long-term impact, or legacy, of an accurate climate prediction? As other scholars have noted, the mass media can aid in the construction and public understanding of symbols of risk that do not always match scientific findings but that may have important social and political outcomes (Krimsky and Plough, 1988). Chapters 5 and 6 both address this issue from the forecast users' perspectives.

Most risk communication—and indeed the science of El Niño—is framed in the terms of probable likelihood, rather than certainty. For example, most daily and seasonal weather forecasts are couched in terms of the probability that something may occur. Journalists who work with these predictions understand that they represent the best estimate of what will happen, rather than certainty, at least in terms of weather reports. News coverage of issues such as the incidence of certain types of cancers or heart disease, other forms of risk, also are framed in such terms. While other studies have indicated that the public tends to discount the probabilistic nature of risk communication, journalistic accounts generally do not, even though there have been well-documented and embarrassing exceptions.

Coverage of El Niño began in this expected way, but in journalistic accounts it became a certitude. This certainty became part of the meta-narrative of the event.

While, on one level, the scientific community should welcome such acceptance of a prediction, the fact that the El Niño–based weather predictions, at least as they were transmitted in the mass media, were so accurate raises other important issues: Will subsequent predictions receive less scrutiny from the popular press and policy elites than warranted? Or, if the public begins to believe that such events as El Niño are harbingers of global change, will that belief fuel other sorts of preparedness and mitigation activity? While an answer to these questions is far beyond the scope of this individual study, media coverage of El Niño does provide important indicators of potential impacts. The most significant of them are as follows:

- Accurate predictions, even though they remain scientifically problematic, will be recounted as fact once an event is under way and outcomes fulfill the predicted events.
- It is possible that issues such as widespread climate change can become part of the news agenda and permeate the popular culture as well. This salience to the public has important political consequences.
- Subsequent predictions, for example the 1998–1999 La Niña phenomenon, may receive less critical journalistic examination because of the perceived accuracy of the preceding El Niño forecasts.
- The totality of the news coverage suggests that mediated risk communication about El Niño may have functioned as a signal event regarding the larger issue of global climate change. This signal influences both public understanding and political activity.
- Agencies that are in the business of issuing forecasts and warnings may have to deal with unintended consequences of success and the resulting fallout when, inevitably, a long-range prediction about future weather turns out to be just that: a prediction.

NOTES

1. Stories for the content analysis were obtained in full text from the Nexis/Lexis data-base. The individual news story served as the unit of analysis. Stories were analyzed for the following: (1) name and data of publication, (2) reporter by-line and specialty, (3) whether the story was hard-news or feature, (4) story length, (5) the generalized subject matter of the individual reports, (6) number and kind of sources cited (particularly scientific sources), (7) the various agencies, when named, that journalists cited as sources for their warnings, (8) whether science was portrayed in the stories as evolving from "fact," (9) the specific benefits and risks attributed to El Niño, (10) whether mitigation information or mention of specific mitigation activity was included in the story, and (11) whether stories included mention of policy responses to El Niño.

Broadcast news accounts were obtained from Burrell's data-base and were analyzed using the same criteria. In addition, placement of a broadcast story within a newspaper was also noted. Intercoder reliability on all items varied from 100 to 81 percent, depending on the item.

REFERENCES

Anderson, A. 1997: *Media, Culture and the Environment*. New Brunswick, NJ: Rutgers University Press.

Carey, J. W. 1989. *Communication and Culture*. Boston: Unwind Hyman.

Graber D., 1988. *Understanding the News*. New York: Longman.

Iynegar, S. 1994. *Is Anyone Responsible?* Chicago: University of Chicago Press.

Krimsby, S., and Plough, A. 1988. *Environmental Hazards: Communicating Risk as a Social Process*. Dover, MA: Auburn House.

Lowery, S., and DeFleur, M. 1986. *Milestones in Mass Communication Research*. New York: Longman.

Shanahan, J. 1993. Television and the cultivation of environmental concern. In *The Mass Media and Environmental Issues*, ed. A. Hansen. Leicester: Leicester University Press.

Signoriellei, N., and M. Morgan (eds.). 1990. *Cultivation Analysis: New Directions in Research*. Beverly Hills, CA: Sage.

Singer, E., and Endreny, P. 1987. Reporting hazards: Their benefits and costs. *Journal of Communication*, 7, 10–16.

Wilkins, L. 1987. *Shared Vulnerability*. Westport, CT: Greenwood Press.

4

 The Scientific Issues Associated
with El Niño 1997–1998

STANLEY A. CHANGNON

The development of a record large El Niño event and its ensuing major effects on the nation's weather over an eight-month period created a scientific event of major proportions. Key science-related questions that developed during El Niño 97–98 included:

- Who was issuing El Niño–based climate predictions and for what conditions?
- What kinds of weather conditions were caused by El Niño?
- What types of impacts were being projected as a result of the El Niño weather?
- How accurate and useful were the El Niño–based climate predictions?
- How accurate were the oceanic predictions relating to the development, intensification, and dissipation of El Niño 97–98?
- Was the record-size event caused by global warming?

Answers to such questions define the scientific information transmitted to the public, the scientific community, and decision makers during the event. This assessment focused on the scientific information that appeared during the period from May 1997 to June 1998, but it also included information that appeared a few months after El Niño ended (i.e., into early 1999), since these issuances reflect the thoughts and findings generated by scientists during the event. Topics assessed included: (1) the sources of the scientific information, (2) how the information was interpreted and by whom, (3) the accuracy of what was presented by different sources, and (4) the scientific issues that emerged, some of which involved disagreements and/or caused potential confusion for decision makers and the public. Most of the information assessed herein was extracted from the Internet, newspaper stories, and scientific documents published during the June 1997–June 1998 period.

What scientific information relating to El Niño 97–98 was measured? We assessed the presentations of the physical descriptions of El Niño and ENSO and the predictions, the predictions based on El Niño conditions of future seasonal climate conditions as well as the resulting physical and societal impacts, the verifications of the seasonal climate predictions, and other, more general information about El Niño 97–98 that emerged, such as its magnitude in comparison to past El Niño events and its possible relationship to other conditions, such as global warming.

The seasonal climate predictions based on El Niño were classified according to whether they pertained to national and/or regional conditions. They were further classified according to three primary forecast conditions: (1) temperature (magnitudes and departures from normal), (2) precipitation, including snowfall (magnitudes and departures from normal), and (3) storms (more or less frequent than normal). Impacts assessed were those based on the predicted climate conditions, and the projected impacts were classified according to sectors such as agriculture, energy, water resources, and economic effects.

The assessment of scientific information focused on that which pertained to conditions in the United States and, as noted, primarily that issued during the fourteen-month study period from May 1, 1997 through June 30, 1998, the time that encompassed the beginning and the end of El Niño 97–98 and its influences on U.S. weather conditions.

El Niño 97–98 had an interesting evolution from its initial detection and the accompanying early predictions to the ensuing growth of public interest and the continuing predictions of seasonal climate conditions, including the impacts expected. These events were intermeshed with the actual weather conditions that occurred across the United States during this amazing fourteen-month period. Thus, considerable attention is given to the temporal distribution of the scientific information generated.

The enormous public interest in and awareness of El Niño 97–98 and the seasonal climate predictions based on the event have the potential for at least four major impacts. These include the effects of the El Niño information on:

- NOAA and its forecasting units, including the Climate Prediction Center;
- The nationwide use of long-range predictions;
- The credibility of the atmospheric and oceanic sciences;
- The use of mitigating measures to reduce weather losses.

The first part of this chapter presents a description of the data and information assessed, including the entities that issued the information and the means by which the information was transmitted. The next section describes and assesses the contents of the scientific information, including the frequency of types of oceanic and climate predictions, the number of impact predictions sorted by sector, and the number of verifications relating to the climate predictions. Special attention is given to the temporal variations found in the scientific material issued during the El Niño 97–98 period. The third section defines and describes the major scientific issues that arose as a result of El Niño 97–98, and the final section summarizes the major findings.

SOURCES OF SCIENTIFIC INFORMATION

Sources of information about El Niño were assessed in two ways: (1) according to the type of institutions/individuals who issued the information, and (2) according to the means used to transmit the information. Institutions and individuals that issued predictive information included federal agencies, regional climate centers, state climatologists, private weather firms, newspaper writers, national laboratories, and universities. Individuals involved included a variety of atmospheric-oceanographic experts, a few nonatmospheric scientists, and many news writers. The institutions assessed as having issued scientific information about El Niño and related U.S. weather conditions during the study period are listed in the appendix to this chapter.

The scientific information assessed was transmitted through the following mediums: (1) newspaper articles, (2) popular magazines, (3) scientific papers and reports, (4) papers and talks presented at conferences and workshops, (5) the Internet, and (6) government agency reports and news releases. Certain advertisements issued during the September 1997–April 1998 period included predictions of weather and serious impacts. For example, a major tire company predicted a bad winter and gave that as a reason one should purchase snow tires. Other advertisements came from investment firms that were advertising that their investments were based on special El Niño forecasts of weather conditions that would hurt the grain markets or energy markets (see chapter 6). Investigations of several advertisements revealed that they were not based on any supportable scientific information, so their contents were not included in the data analyzed.

The scientific information contained in 279 separate articles about El Niño 97–98 issued between May 1, 1997, and June 30, 1998, was examined and classified. The number and sources of articles assessed appears in Table 4-1,

TABLE 4-1.   Means used to transmit information about El Niño 97–98 assessed for scientific content and issued during the May 1, 1997-June 30, 1998 period.

Means	Number
Internet items	95
News articles	150
Scientific papers	6
Scientific reports	15
Conference/workshop reports	3
Popular magazine articles	10

which shows that the preponderance of the information, as expected, came from newspapers and the Internet, with their contributions totaling 245 of the 279 sources. Figure 4-1 illustrates a few of the hundreds of offerings on the Internet.

Nine of the 279 articles were from foreign sources and did not treat U.S. conditions specifically. The remaining 270 articles that were U.S.-specific were analyzed for their scientific content and predictions. Contents were examined for each of the following topics:

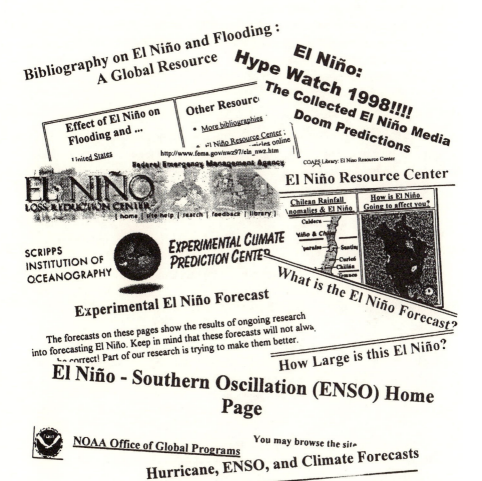

FIGURE 4-1. Examples of headlines on El Niño found on the Internet during the El Niño period, from September 1997 to June 1998.

- Description of El Niño 97–98
- Description or reference to 82–83 El Niño
- Prediction of oceanic conditions
- Prediction of U.S. climate conditions, nationally and/or regionally, and what conditions were predicted (temperature, precipitation, and/or storms)
- Prediction of U.S. impacts, classified by sectors, including hydrologic, environmental, agricultural, economic, societal, energy, health and welfare, and recreation
- Issuance of verifications of the climate predictions
- Issuance of general scientific information such as sources of further information about El Niño
- Raising of scientific issues related to El Niño, such as its causes, including global warming.

The originators of the articles were also classified, first, as to whether the article was authored by one or more individuals or was from an institution without any individual cited. These author/institutional originators were classified as federal agencies, state or regional agencies, universities, national laboratories and scientific institutions, private weather companies, or newspapers or magazines.

The number of articles assessed for each source appears in Table 4-2. There were 100 articles from federal agencies, and forty-three of these identified one or more scientists or administrators who authored the article. The other fifty-seven federal agency articles were simply institutional releases without any named person. Newspapers and popular magazine articles authored by news writers totaled eighty-one, and thirty-six articles had no individual or individuals cited as author. Forty-five articles had identified authors, and, as we discuss later, most of these were news or science writers. Most of these stories, forty-three in all, contained supporting scientific information in the form of statements or quotations from one or more scientists used as "sources". Most articles from private weather companies, state/regional agencies, and universities were authored

TABLE 4-2. Number of articles about El Niño 97–98 issued by institutions and/or publications with and described as authored by individuals or by the releasing institution.

Source	Institution without a named author	Author or authors named
Federal agency	57	43
State or regional agency	5	10
University	5	24
National laboratory	19	12
Private company	3	11
Newspaper/magazine	36	45

by scientists, whereas many of the articles from national laboratories did not cite a scientist. Interpretation of the scientific information expressed in the eighty-one articles authored by a newspaper or a news writer should recognize that a nonscientist presented the information, which may or may not be correct.

CONTENTS OF SCIENTIFIC ARTICLES
Predictions of Weather Conditions

A wide variety of institutions and individuals issued weather predictions based on El Niño's influence on the atmosphere over the United States. The term predictions, as used here, has the same meaning as forecast and outlook. The predictions concerned primarily the mean states of future seasonal climate conditions (i.e., above, near, or below normal), and they most commonly addressed expected temperature levels, precipitation levels, and/or degree of storminess. Some were state-specific, but most were regional in nature. That is, "the winter temperatures in the northern United States are predicted to be above normal." Another set of forecasts addressed future oceanic conditions, including the SSTs expected in various parts of the Pacific Ocean.

Predictions were issued routinely or intermittently by several institutions listed in the Appendix, including the Climate Prediction Center (the official U.S. predictions), COLA, Scripps, the Florida State group, IRI, and the WMO. CPC, through its Experimental Long-Lead Forecast Bulletin (which appeared on the Internet) issued on a quarterly basis forecasts of oceanic conditions as determined by fifteen different models in use at CPC and at several other institutions. Some institutions issued occasional oceanic and seasonal climate predictions; among these were JPL, NCAR, NASA Goddard, CIRES, PMEL, Midwestern Climate Center, and Western Regional Climate Center. Infrequent, typically one-time-only predictions, such as the forecast of the magnitude of the spring tornado season or winter storms (Figure 4-2), were issued by a few federal agencies and by individuals from state agencies, in the private sector, or on staff at various universities. These totaled nineteen during the July 1997–May 1998 period. The total number of institutions and individuals that issued one or more climate predictions on the basis of El Niño conditions was thirty-two. Figure 4-3, which is based on titles of a few of the scientific papers and reports issued during El Niño, illustrates the diversity of the subject matter presented.

Table 4-3 presents the distribution of the number of climate predictions by area and type, sorted according to the sources of the articles. Of course, not all of the 270 articles included predictions, on a regional or national scale, about future conditions due to El Niño 97–98. During June–August 1997 there were sixteen (of twenty-five articles assessed) with climate predictions; thirty-seven (of seventy-three) in September–November, thirty-seven in December–February (of 100), and twenty (of sixty-six) in March–May 1998 contained some form of long-range climate prediction. Note the temporal decline in predictions, given

FIGURE 4-2.   Heavy ice damage to trees lining a New Hampshire country road after a major January 1998 ice storm that was attributed to El Niño. The storm caused $400,000 in losses in New England. (Courtesy Bob Melville)

the total number of articles issued, over time, going from about 60 percent in the summer of 1997 to less than 30 percent by March–May 1998.

Inspection of the issuances from federal agencies shows that about a third of the 100 articles issued addressed regional and/or national climate predictions, with slightly more addressing future precipitation than temperatures or storm conditions. For obvious reasons, the state and regional institutions focused their predictions totally on regional conditions, with a great emphasis on temperature and precipitation conditions and less attention to storms. The storm conditions addressed most often were hurricanes, coastal storms, and tornadoes.

TABLE 4-3.   The number of times articles carried predictions of national and/or regional climate conditions, as issued between May 1, 1997, and June 30, 1998, sorted according to sources. The numbers in parentheses show the percentage of the total articles issued by each source.

Source	National predictions	Regional predictions	Precipitation predictions	Temperature predictions	Storminess predictions
Federal agency	30 (30%)	34 (34%)	34 (34%)	29 (29%)	25 (25%)
State or regional agency	0 (0%)	9 (60%)	9 (60%)	7 (47%)	3 (20%)
University	11 (38%)	11 (38%)	8 (28%)	5 (17%)	15 (52%)
National laboratory	10 (32%)	8 (26%)	11 (35%)	9 (29%)	6 (19%)
Private company	2 (14%)	7 (50%)	4 (29%)	4 (29%)	4 (29%)
Newspaper/magazine	18 (22%)	18 (22%)	19 (23%)	14 (17%)	19 (23%)

SPECIAL CLIMATE SUMMARY 97/3
NOVEMBER 1997
CLIMATE CONDITIONS ASSOCIATED WITH
THE 1997–1998 EL NIÑO:
IMPACTS AND OUTLOOK

REPORTS
El Niño and Climate Prediction

El Niño forecast fails to convince sceptics

[LONDON] Predictions that the expected El
Niño — the appearance of warm waters off
the coast of South A---

Greenhouse Warming, Decadal Variability, or El Niño?
An Attempt to Understand the Anomalous 1990s

The 1997 El Niño/
Southern Oscillatio
(ENSO 97-98)

GEOPHYSICAL RESEARCH LETTERS,
El Niño and climate change

THE 1997-98 EL NIÑO

- one of the most severe
ENSO events in history

Possible Impacts on the Property Insurance Industry

NETWORK

NEWSLETTER

Anomalous ENSO Occurrences: An Alternate View*

Interdecadal Modulation
of ENSO Teleconnections

Volume 13, Number 4

Editorial
The El Niño Olympics; or
The Search for the El Niño of th--

Forecasts of Tropical Pacific SST Using a Comprehensive Coupled
Ocean-Atmosphere Dynamical Model

--d by Tony Barnston², Ming Ji¹, Arun Kumar¹ and Ants Leetmaa²

The effect of El Niño on U.S. landfalling hurricanes.

Experimental Long-Lead

FORECAST

Spaceborne Sensors Observe
El Niño's Effects on Ocean
and Atmosphere in North Pacific

Tropical Cyclone Occurrences in the Vicinity of Hawaii: Are the Differences between
El Niño and Non–El Niño Years Significant?*

FIGURE 4-3.   Titles of a few of the many scientific papers and reports published
during El Niño 97–98, reflecting the diversity of the topics treated.

Table 4-3 also presents percentages of the total issuances for each source to allow comparison of the sources and their areas of interest. These figures show, for example, that the university-issued climate predictions were more often focused on storm conditions (52%) than were predictions issued by the other sources. The percentages also reveal that the newspaper-authored articles typically carried less information on predictions of any specific condition (23% or less) than did the articles from all other sources. The predictions of the private weather companies focused on regional precipitation conditions to a greater degree than did those of other sources, likely reflecting their customers' concerns.

Predictions of Oceanic Conditions

During the May 1, 1997–June 30, 1998 period, seventy-five of the 270 articles issued carried predictions of the future oceanic conditions in the Pacific Ocean. Most of these were found in federal articles (forty total) and in releases from national laboratory sources (twenty-one). These two types of institutions operate climate models capable of such predictions. After May 1997, little disagreement was found in the contents of these predictions. The earlier forecasts (pre-May 1997) of the onset and intensity of El Niño 97–98 were not as accurate. Once El Niño developed in May–June 1997, most climate models accurately predicted its peak in 1997–1998, but several models failed to predict its sudden dissipation during May–June 1998.

Predictions of Impacts from El Niño–Generated Weather Conditions

The distribution of releases containing predictions of impacts expected from predicted El Niño–generated climate conditions (Table 4-4) reveals that the most frequently predicted conditions were economic or hydrologic (usually flooding) effects, followed by agricultural and environmental impacts. The sectoral analysis shows that the newspaper-authored articles focused more frequently on expected economic, agricultural, and recreational impacts, whereas releases from the federal agencies most often featured predictions of hydrologic conditions. Forty-four percent of the 100 federal agency issuances contained some form of impact prediction, whereas 60 percent of the newspaper-authored articles addressed expected impacts from El Niño–driven weather conditions (see chapter 3). Eight of the fifteen articles issued by state and regional agencies addressed impacts, and most of these addressed water resource-flooding problems. The fourteen articles issued by scientists at private weather companies all predicted impacts, and most focused on a specific region, such as the Midwest, Northeast, or West Coast.

TABLE 4-4. Number of times articles included predictions of weather-related impacts in various sectors, as issued between May 1, 1997, and June 30, 1998, sorted according to sources.

Source	Environment	Water	Agriculture	Economy	Energy	Recreation	Society	Health
Federal agency	12	26	8	16	7	2	5	13
State or regional agency	0	6	2	1	1	0	1	1
University	5	5	6	3	0	2	0	1
National Laboratory	6	9	2	4	0	0	0	1
Private company	1	0	5	3	3	1	0	0
Newspaper/ magazine	11	9	19	29	4	10	8	7
Totals	35	55	42	56	15	15	14	26

Verification of the Climate Predictions

An interesting aspect of the El Niño–based seasonal predictions involved efforts to verify their correctness. About 10 percent of the 270 articles contained information assessing the accuracy of all or some form of the predicted conditions. Federal agencies issued fifteen verifications, scientists in regional agencies issued four verifications, newspaper authors issued three, national laboratories issued four, and two came from university scientists. Most claimed high accuracy for the seasonal predictions. The state-regional articles focused on regional conditions (e.g., Midwestern snowfall, California storms), whereas the federal issuances focused on oceanic or seasonal weather predictions for large areas across the nation.

The National Oceanic and Atmospheric Administration (NOAA) was the leading issuer of verifications, with articles from the Secretary of Commerce, the NOAA Administrator, the heads of the National Weather Service (NWS), the Office of Global Programs (OGP), and the Climate Prediction Center (CPC), and numerous staff members in field offices of the NWS. Figure 4-4 shows the 103-year history of winter (December–February) temperatures in the Northeast (forecast to be above normal in 1997–1998), and for winter precipitation in the Southeast (forecast to be above normal in 1997–1998). As the figure shows, winter values in 1997–1998 met their forecast values and were at near-record levels.

Temporal Distributions of Predictions

The focus of the various types of predictions shifted with time during the study period. The number of articles assessed for scientific content during each month is shown in Table 4-5. This reveals a rapid growth in the total number of articles during the summer of 1997.

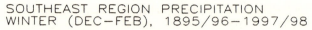

SOUTHEAST REGION PRECIPITATION
WINTER (DEC-FEB), 1895/96-1997/98

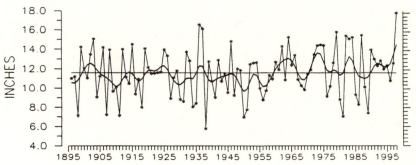

NORTHEAST REGION TEMPERATURE
WINTER (DEC-FEB), 1895/96-1997/98

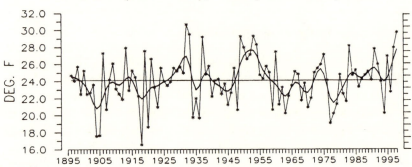

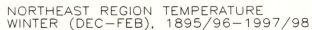

FIGURE 4-4.   Graphs depicting winter (December–February) values for precipitation in the Southeast United States (top graph) and temperatures in the Northeast United States (lower graph), for the past 103 years. These reveal record high precipitation during the 1997–1998 El Niño winter in the Southeast and also reveal that the 1997–1998 winter was the Northeast's second warmest winter on record. (The two areas are shown on Figure 1-3. (National Climatic Data Center)

More than half (146) of the 270 articles contained some form of impact prediction, and fewer than half, 112 (41%), offered climate predictions. A fourth of the articles contained predictions of oceanic conditions.

A considerable number of the 270 articles (fifty-four) presented various types of scientific information related to El Niño. This "general information" category was wide ranging, with some articles describing the ENSO processes; others listing El Niño data sources or scientists with expertise; others addressing issues such as the relationship of global warming to El Niño 97–98 or the relative severity of the event; and others discussing forecast techniques or reasons for the success or failure of the predictions. Assessment of the articles that contained a reasonably thorough description of El Niño and ENSO revealed that eighty-four, or 31 percent, discussed the physical processes involved and how El Niño had

TABLE 4-5. Temporal distribution of the number of articles containing scientific information about El Niño 97–98 and assessed for predictions and other scientific information.

Month	Total no. of articles	Ocean predictions	Climate predictions	Impact predictions	Climate verification	General information
May 1997	3	3	0	1	0	0
June	5	2	3	2	0	1
July	7	3	4	2	0	2
August	13	5	10	8	0	2
September	21	6	9	7	1	9
October	27	3	16	16	2	3
November	25	6	16	15	1	4
December	22	6	11	8	1	4
January 1998	37	16	15	17	4	6
February	31	5	8	20	7	5
March	32	5	9	18	7	7
April	27	11	8	18	3	8
May	7	1	2	5	1	1
June	13	3	1	9	1	2
Totals	270	75	112	146	28	54

affected U.S. weather conditions. As expected, issuance of these descriptions was relatively more frequent in the June–September 1997 period than in the following months, as explained in chapter 3.

The temporal distributions of the climate and impact predictions (Table 4-5) are revealing. The number of articles predicting future weather conditions exceeded those predicting impacts from May 1997 through December 1997, but thereafter the monthly numbers of predictions of impacts exceeded the number of climate predictions, and the difference increased with time.

The number of predictions of tropical Pacific Ocean conditions increased dramatically early in 1998 as concern and interest grew over the expected conditions in the spring, summer, and ensuing seasons—was El Niño to continue, or would La Niña prevail? As expected, articles addressing the verification of the El Niño predictions grew in number as the 1997–1998 winter ended, an appropriate time to evaluate the winter predictions. The earlier-issued evaluation articles (during September–December) focused on the predictions of the onset of El Niño 97–98; later ones focused on the advent of the early coastal storms and the severe convective storms as proof of the accuracy of the predictions. Verifications issued in the winter and spring of 1998 focused on the correctness of the winter predictions.

The greatest monthly number of articles (>30 per month) came during January–March 1998. This reflected (1) attention to the success in the fall-winter season predictions, (2) the significant public interest in the topic, and (3) the news media's attention to the issue (see chapter 3). As El Niño 97–98 faded late in the spring, the number of science-oriented articles diminished rapidly.

The temporal variations in the contents of the official ENSO predictions (advisories) issued by the Climate Prediction Center (CPC), as illustrated in Figure 4-5, are discussed in chapter 1. These reveal the important temporal shifts in the status of ocean conditions and the official climate predictions.

Another interesting temporal analysis dealt with the use of scientists as sources of information for input to newspaper articles about El Niño. Table 4-6 shows the time distribution of news articles that identified one or more scientists as having helped develop the story material and of those that were authored without any recognition of input from a scientist. The distributions show marked temporal shifts. In the fall of 1997, twenty-two of the thirty-four articles were based on input from scientists, who were named and information presented credited to them.

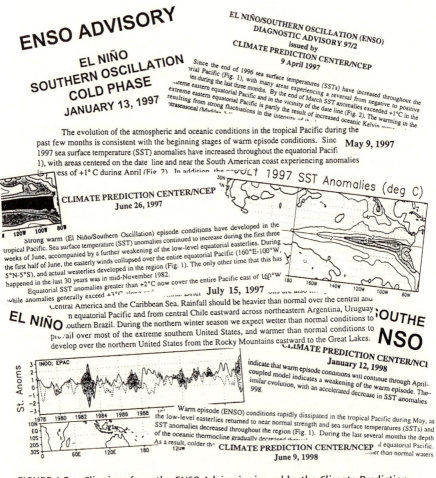

FIGURE 4-5. Clippings from the ENSO Advisories issued by the Climate Prediction Center during 1997–1998. These illustrate some of the major changes that were observed and predicted.

TABLE 4-6. The temporal distributions of the number of newspaper articles issued with and without use or naming of scientists as sources of information for the material presented in the articles.

Period	Scientists cited/ acknowledged	No citation of scientists in the article
September–November 1997	22	12
December–January 1998	9	15
February–March	3	7
April–June	13	7

This trend shifted during December–March, when most newspaper articles containing scientific information failed to indicate a scientific source. This shift suggests that news writers had become sufficiently confident in their knowledge of the conditions to write about El Niño without any reported evidence of having consulted a scientist. And, as noted in chapter 3, the stories had shifted from explaining El Niño and its predictions to blaming El Niño for various weather events and related damages. This trend ended during April–June, when the number of articles identifying scientists as sources increased again. This may have been a result of the shifting El Niño conditions and the differing predictions being issued by various sources in the spring, creating a situation in which the news writers again perceived a need to use scientists as consultants for their stories. That is, the sense of scientific certainty about El Niño effects had diminished among news writers, and the focus of the El Niño stories had changed.

SCIENTIFIC ISSUES RAISED BY EL NIÑO

The contents of the 270 articles assessed revealed the emergence of five scientific issues that gained wide attention during the May 1997–June 1998 period. Some centered on scientific differences and uncertainties. In this section, descriptions of these issues are presented, generally as they evolved over time. The issues defined were: (1) weather events attributed to El Niño, (2) questions and confusion about the climate predictions based on El Niño, (3) rating of the magnitude of El Niño 97–98, (4) the relationship between global warming and El Niño 97–98, and (5) differences in predicted impacts and the reactions of the public and decision makers. Each of these issues is discussed in this section. A sixth part of this section analyzes the El Niño–related topics presented at two major scientific conferences held at the end of 1998. These reflect how the scientific community reacted to El Niño 97–98 and the issues it addressed.

Attribution of Major Weather Events to El Niño

During El Niño 97–98 there were numerous notable storms, and various views were offered by various scientists about whether these events were totally due to

El Niño's influence, partially due to El Niño, or not related to El Niño's influence on the atmosphere. Events attributed totally or partially to El Niño's influences on U.S. weather patterns involved four storm types: hurricanes, winter storms, coastal and inland rain-wind storms, and tornadoes. In general, the seasonal temperature levels and precipitation amounts were identified as El Niño–related without any questions being raised about whether this attribution was valid.

The newspaper headlines shown in Figure 4-6 illustrate the diversity of views expressed about El Niño's influence on various weather conditions. The greatest differences of views about attribution focused on various tornado occurrences during December–April. A National Weather Service (NWS) scientist proclaimed in February, "You can blame a great deal of what is going on in the U.S. since the fall on El Niño" (*Cleveland Plain Dealer*, February 2, 1998). The question became, What exactly could not be attributed to El Niño's influence?

Hurricane activity was considered strongly influenced in two distinctly different ways. First, the lack of Atlantic hurricanes during the summer and fall of 1997 was widely proclaimed as a result of the influence of El Niño (Gray, November 29, 1997; Douglas, September 2, 1997). This was correctly viewed as a huge blessing (*U.S. Water News*, January 1998). During September–October three large hurricanes (Linda, Nora, and Pauline) developed in the Pacific just west of Mexico and caused great damage in Mexico, including 200 deaths, and two produced heavy rains in southern California. Some meteorological analysts blamed these hurricanes on El Niño (*Chicago Tribune*, September 14 and 25, 1997), and a NWS hurricane forecaster indicated that the strength of these hurricanes was enhanced by El Niño (Price, October 9, 1997). The Federal Emergency Management Agency (FEMA) further claimed that the Pacific hurricanes were produced by El Niño (FEMA, September 29, 1997).

Importantly, the advent of these strong Pacific hurricanes and an October snowstorm in the High Plains, coupled with the lack of Atlantic hurricanes, collectively helped convince the news media and the public that El Niño's influence on U.S. weather was both real and strong. As shown in chapter 3, after October the media blamed every weather event on El Niño's influence.

Most areas of the northern United States had few winter storms and snowfall levels well below normal during the winter of 1997–1998 (see Figure 2-14). However, two major winter storms occurred, and both were attributed to El Niño by some scientists. The first was a strong October snowstorm that crossed the High Plains, wreaking great damage, and a NOAA scientist said the storm was strengthened by El Niño (*USA Today*, October 28, 1997). The storm's unusual intensity and early appearance were factors that further directed the public's attention to El Niño, as explained in chapter 3. Chapter 7 addresses the various views expressed by scientists about El Niño and its relationship to the October snowstorm. Many months later, a scientist claimed that to consider the huge winter storm as related to El Niño was "ridiculous," but he also stated that the occurrence of a higher-than-normal number of storms when El Niño occurred

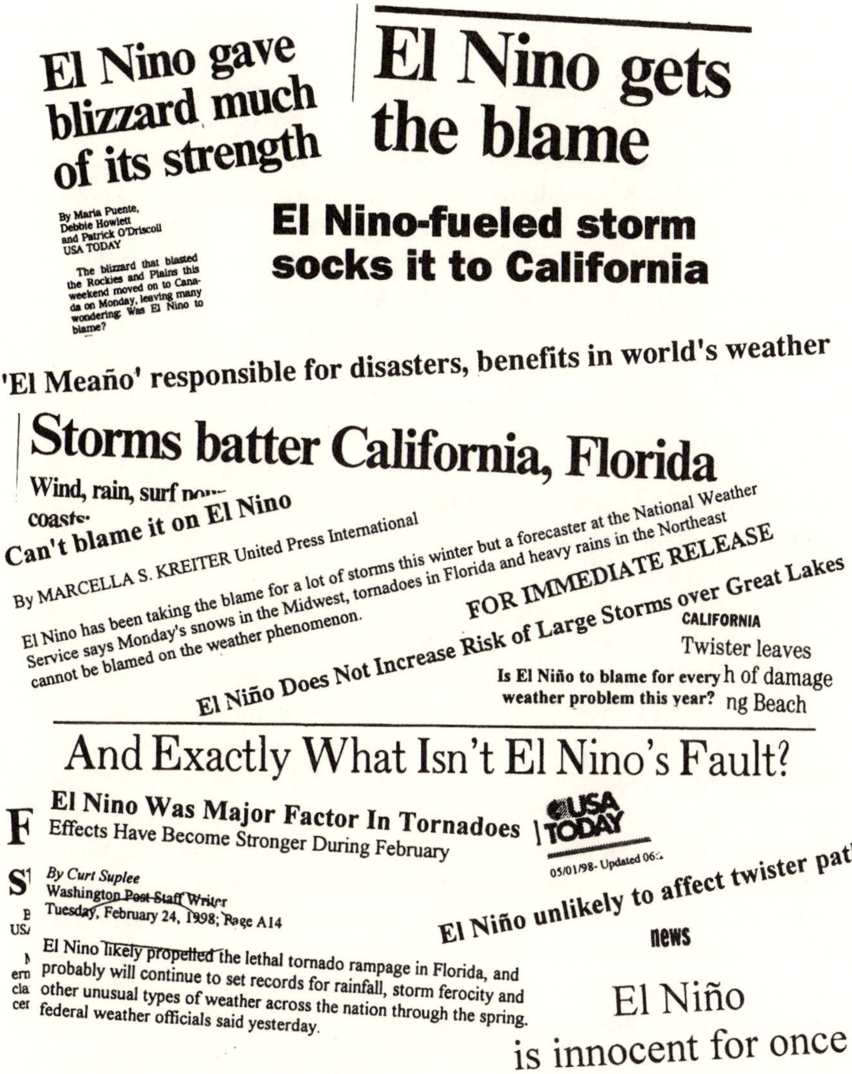

El Nino gave blizzard much of its strength

By Maria Puente, Debbie Howlett and Patrick O'Driscoll USA TODAY

The blizzard that blasted the Rockies and Plains this weekend moved on to Canada on Monday, leaving many wondering. Was El Nino to blame?

El Nino gets the blame

El Nino-fueled storm socks it to California

'El Meaño' responsible for disasters, benefits in world's weather

Storms batter California, Florida

Wind, rain, surf n~~
coasts.
Can't blame it on El Nino

By MARCELLA S. KREITER United Press International

El Nino has been taking the blame for a lot of storms this winter but a forecaster at the National Weather Service says Monday's snows in the Midwest, tornadoes in Florida and heavy rains in the Northeast cannot be blamed on the weather phenomenon.

FOR IMMEDIATE RELEASE

El Niño Does Not Increase Risk of Large Storms over Great Lakes

CALIFORNIA

Twister leaves

Is El Niño to blame for every h of damage
weather problem this year? ng Beach

And Exactly What Isn't El Nino's Fault?

El Nino Was Major Factor In Tornadoes
Effects Have Become Stronger During February

USA TODAY

05/01/98- Updated 06:.

El Niño unlikely to affect twister path:

news

By Curt Suplee
Washington Post Staff Writer
Tuesday, February 24, 1998; Page A14

El Nino likely propelled the lethal tornado rampage in Florida, and probably will continue to set records for rainfall, storm ferocity and other unusual types of weather across the nation through the spring, federal weather officials said yesterday.

El Niño
is innocent for once

FIGURE 4-6. There was considerable scientific uncertainty about which weather events in North America during 1997–1998 were caused by El Niño. The media and many scientists tended to blame almost everything on the event, particularly after October 1997.

was "expected" (*Detroit News*, May 7, 1998). If more are expected, which ones are due to El Niño, and which ones are not? No wonder the public was confused about what weather conditions to blame on El Niño! It is easy to understand why major newspapers ran headlines like this one: "El Niño disrupts the weather and confounds scientists: The debate over effects of El Niño rages unabated" (*Minneapolis Star Tribune*, February 27, 1998).

Early in January 1998, an enormous ice storm developed in the northeastern United States and Canada and created huge losses, $5 billion in Canada and $400 million in New England (Figure 4-2). This record storm also was blamed on El Niño (Associated Press, January 11, 1998).

The official predictions of climate conditions related to El Niño included warnings of a higher than usual number of fall and winter coastal storms along the West Coast (CPC, August 4, 1997). An early December coastal storm struck California and was labeled the "first" El Niño coastal storm (Associated Press, December 7, 1997). As shown in Figure 4-7, all subsequent coastal rain and wind storms on the east and west coasts during the winter were also attributed to El Niño (*USA Today*, February 3, 1998). In addition, numerous storms producing heavy rains inland were also attributed to El Niño. A mid-December storm in Texas produced eight inches of rain and six inches of snow and led to deaths of eight persons, and a private–sector meteorologist proclaimed that the storm was a result of El Niño (*USA Today*, December 22, 1997). A series of flood-producing storms during January–March in various parts of the Deep South was attributed to El Niño (*Champaign-Urbana News Gazette*, March 15, 1997).

Tornadoes became the fourth type of storm caught up in the debate over what weather conditions were caused by El Niño. Wide differences of opinion were stated, and one scientist summed up the situation by saying, "Climatologists and forecasters disagree over just how much to blame El Niño for the tornadoes across Florida and the Gulf States" (*Minneapolis Star Tribune*, February 27, 1998). Seven tornadoes in Florida on February 2 were attributed to El Niño (*USA Today*, February 3, 1998). Then, a major tornado outbreak on February 22–23 in Florida, which killed forty-two persons, was attributed to El Niño by NWS scientists (*Washington Post*, February 24, 1998) and by news writers (*Champaign-Urbana News Gazette*, March 1, 1998). Another NWS meteorologist disagreed, stating that the Florida tornadoes were only "partially" caused by El Niño (*Chicago Tribune*, February 23, 1998).

On February 28, the head of NWS's severe storm forecast center proclaimed that there was no correlation between El Niño and the tornadoes that had occurred (*USA Today*, February 28, 1998). Subsequent news stories early in March offered conflicting views on El Niño's role in causing tornadoes. One NWS storm analyst stated that the tornadoes and winters storms should *not* be attributed to El Niño (UPI, March 9, 1998), and a day later another NWS scientist listed the various weather events and storms that were caused by El Niño (*International News*, March 10, 1998). After tornadoes late in March killed two persons in Minnesota, the local NWS forecaster cited El Niño as the cause of the twisters (Powers, March 31, 1998). A group of scientists considered to be tornado experts offered conflicting views about El Niño's influence on storm activity in a national news story released early in April (*USA Today*, April 3, 1998).

In March the debate shifted to what would happen in the near future when the normal "tornado season" (April–May) occurred. One university scientist

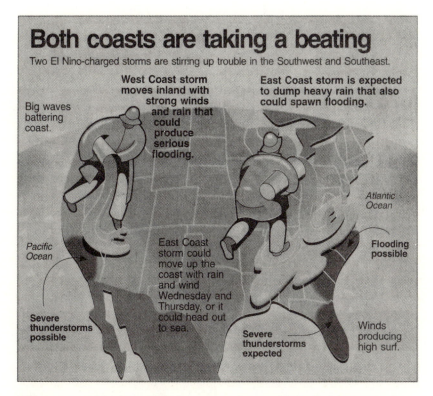

Both coasts are taking a beating

Two El Nino-charged storms are stirring up trouble in the Southwest and Southeast.

West Coast storm moves inland with strong winds and rain that could produce serious flooding.

East Coast storm is expected to dump heavy rain that also could spawn flooding.

Big waves battering coast.

Atlantic Ocean

Pacific Ocean

East Coast storm could move up the coast with rain and wind Wednesday and Thursday, or it could head out to sea.

Flooding possible

Severe thunderstorms possible

Severe thunderstorms expected

Winds producing high surf.

FIGURE 4-7.   Major winter storms on both coasts occurred during 1998, and, as shown here, they were attributed by the press to El Niño. (Copyright 1998, *USA TODAY*; reprinted with permission)

said that since El Niño had shifted the jet stream, it would shift tornado locations from their normal places. Other university scientists noted as experts on tornadoes in the High Plains agreed that El Niño would lead to shifts in tornado locations, without stating where they might occur (*Chicago Tribune*, April 13, 1998). Another scientist predicted that fewer tornadoes would occur in the Midwest, and the Iowa state climatologist predicted fewer tornadoes in Iowa in the spring (*Des Moines Register*, April 6, 1998). The head of the NWS severe storm forecast center proclaimed that no capability existed that would allow scientists to predict tornado activity in the coming tornado season (*USA Today*, February 28, 1998). A later assessment of the 1998 weather conditions in the United States tied the 1998 tornadoes and floods in the southeastern United States to El Niño's influence (Le Comte, 1999). Ultimately, Illinois experienced 106 tornadoes in 1998, more than any other state, and Illinois is not in the nation's tornado belt, suggesting that there was a major shift in the nation's tornadic activity.

Ants Leetma, the director of the CPC, announced late in March that all eighteen storms between October 1997 and March 1998 that caused enough damage for the regions affected to be declared disaster areas by the president were partially due to El Niño (*USA Today*, March 25, 1998). A scientist with a long background in ENSO research summed up the unresolved debates over attribution by saying that "not everything being blamed on El Niño was a result of El Niño," and he claimed that blaming every day weather events was "wrong . . . except in a few cases such as the West Coast being slammed by storms" (*Detroit News*, May 7, 1998). A confusing situation at best!

Scientists during El Niño 97–98 revealed great uncertainty as to what could be attributed to El Niño. Most scientists were willing to attribute changes in atmospheric circulation patterns over the United States and large, synoptic-scale storms to the event, but many remained unsure as to whether to attribute smaller-scale storms, particularly those related to strong, localized convective activity.

Two unexpected atmospheric problems that occurred after the El Niño influence on North American weather had weakened were explained as being related to El Niño. In May and June 1998, large-scale fires in Mexico caused by a severe drought blamed on El Niño led to enormous plumes of smoke that spread over the southern United States and affected air quality (*Champaign-Urbana News Gazette*, June 7, 1998). Late in the spring, the weather became very dry across Florida, and, by June, large fires developed in parts of Florida, requiring evacuations of hundreds, causing crops and homes to burn, and blocking rail and highway transportation arteries. These fires were explained as El Niño–related because the winter storms caused by El Niño had produced near-record November–February rainfall (*USA Today*, June 8, 1998), leading to a profusion of underbrush, which, after the unusual spring dryness, became fodder for the fires (*USA Today*, June 26, 1998). What was not offered was an explanation of the abrupt dryness that set in during the spring in Florida when and where El Niño predictions still called for wet conditions. This is thus an example of an area where the spring precipitation predictions were incorrect.

Questions, Differences, and Confusion
Surrounding the Climate Predictions

Climate outlooks, or predictions, normally arouse scientific debates about their accuracy, and the El Niño–based long-range forecasts were subject to some questioning. The first official predictions about the climate conditions to be expected from the rapidly developing El Niño were issued in June 1997, and by July the official predictions were indicating the strength of the developing event and the kinds of fall and winter climate conditions to expect in the United States and elsewhere (CPC, August 4, 1997).

Several scientists questioned the accuracy of the initial El Niño seasonal forecasts (*Nature*, July 10, 1997). Questions were raised in August about the correctness of the official predictions about El Niño, about whether El Nino would really occur and, if it did, about its nature (*U.S. Water News*, August 1997). In October a university scientist openly challenged the accuracy of the forecasts (*Southern Illinoisan*, October 26, 1997). Another university scientist issued a prediction early in September that called for a severe and stormy winter in the Midwest and noted that his view totally disagreed with the official predictions for winter conditions in the Midwest (*St. Louis Post-Dispatch*, September 4, 1997). Such differences were one factor among several that helped cause many decision makers to decide not to use the seasonal predictions, as discussed in chapter 5.

Early in November, a scientist at the Midwestern Climate Center predicted that the El Niño conditions would lessen the frequency and intensity of winter storms on the Great Lakes (University of Illinois, November 4, 1997). At the same time, university scientists in Michigan announced that the El Niño winter would bring the opposite outcome—worse-than-normal storms on the Great Lakes, with extensive shore damages around the lakes (University of Michigan, October 24, 1997). An article in a national magazine questioned NWS claims about its predictions and statements attributing certain weather events to El Niño (Grenci, 1997). However, most questioning of the official climate predictions during 1997 was minor and infrequent.

Until early in the winter of 1997–1998, the official seasonal predictions being issued by the CPC, as well as those issued by numerous research institutions, were in close agreement about the expected fall and winter weather conditions across the United States. This is no surprise since relatively good forecasts five to six months in advance of a strong El Niño are to be expected because of well-established teleconnections between El Niño influences and the climate conditions in many areas of the United States.

However, in December 1997, a research institution issued an outlook for a cool spring, a different outcome from the official forecast, which called for warm conditions (COLA, December 1997). This prediction opened a period when predictions issued by various institutions for the spring, summer, and fall conditions of 1998 varied considerably. This raised questions among the news media about the accuracy and utility of the new predictions. Figure 3-1 illustrates some of the differences, as expressed in newspaper headlines. The CPC issued information in February announcing that various summer 1998 predictions were quite different, stating, "there is substantial disagreements between the various models over Pacific conditions beginning in mid 1998" (CPC, February 12, 1998). The CPC also stated its expectation that El Niño conditions would continue into spring and early summer (*USA Today*, February 19, 1998). Summer predictions issued by other groups varied; some called for normal summer weather, others for a cool summer, and still others for hot and dry conditions (Douglas, March 7, 1998).

Disagreements among the model-based projections became a frequent topic addressed by the media. Predictions for summer and fall 1998 conditions issued by two agencies, NASA Goddard and the Scripps Institute, were highlighted by the television media as being in disagreement (NBC, February 15, 1998). In March, scientists at a national laboratory questioned the accuracy of all the El Niño–based seasonal predictions because of the major differences in their forecasts for 1998 (*San Jose Mercury News*, March 16, 1998). NASA scientists predicted in March the development of La Niña conditions during the fall of 1998 and its attendant drought-like weather conditions (CNN, March 26, 1998). The official model-based predictions issued at the start of April by the NWS and those of several other institutions (IRI, COLA, COAPS, and Australia's Bureau of Meteorology) contained major differences. An NCAR scientist noted that the media and certain scientists had all issued dire predictions for El Niño's effects and noted that, since many of the actual outcomes were far from being bad, the forecasts had created misleading information and poor understanding about ENSO (Glantz, August 1998). Chapter 7 explores the confusion surrounding the use and misuse of the predictions.

Rating the Magnitude of El Niño 1997–1998

The onset of some warming in the tropical Pacific was first predicted in December 1996 by several climate models. However, none of the forecasts issued during the first part of 1997 captured the strength or the rate of the warming that ultimately occurred in the Pacific (Trenberth, 1999). The oceanic predictions about El Niño's development issued during June–August 1997 kept noting that the area and volume of the warmer waters were immense and likely to become greater than in any past El Niño (*Des Moines Register*, September 16, 1997). El Niño developed so rapidly that, for each month from June to December 1997, a new monthly record high was set for SST values in the eastern equatorial Pacific (McPhaden, 1999). Pronouncements about the rapid growth to record or near-record sizes led to repeated interpretations in the media that El Niño 97–98 was greater than the record-damaging El Niño 82–83 event, making the 1997–1998 event the greatest in 100 years, or the "worst on record" (*Chicago Tribune*, October 14, 1997).

The record oceanic dimensions of El Niño 97–98 were interpreted by some in the news media as creating "the most destructive weather pattern in a century" (*Cincinnati Inquirer*, October 12, 1997). Many news stories noted that major weather disasters would occur globally. Pronouncements about El Niño 97–98's being the "worst ever" continued during the fall season (*St. Louis Post-Dispatch*, November 22, 1997), effectively resulting in a theme of a "gloom and doom" outcome globally and especially in the United States. FEMA's news releases in August and September 1997 emphasized the great strength and size of El Niño 97–98 and indicated that it would match or surpass the extremely damaging El Niño 82–83 (FEMA, August 12 and September 12, 1997).

Some scientists speculated that the El Niño's record dimensions were tied to global warming (*Washington Post*, February 26, 1998), and this led to newspaper cartoons like that in Figure 4-8. The record warm and dry winter in the Midwest was seen as the result of the record El Niño (*Cleveland Plain Dealer*, March 1, 1998). A University of Michigan atmospheric scientist stated, "We have never seen anything like this. This is the strongest El Niño in more than 100 years" (*Detroit News*, March 20, 1998). Others questioned the quality of the historical oceanic data and therefore scientists' ability to make claims of a "once-in-a-100-year-event".

A leading El Niño expert announced that El Niño 97–98 was an order of magnitude greater than El Niño 82–83 (Glantz, March 1998). Later, and with a more philosophical view, the same scientist assessed the various ways of rating El Niño events and explained how the various past events had all been different (Glantz, August 1998). For example, El Niño 97–98 may have had the largest oceanic warming experienced, but its negative impacts in the United States, as reflected in lives lost and dollar losses, were less than those in 1982–1983 (see chapter 6).

Relationship between El Niño and Global Warming

The potential relationship of global warming, caused by an enhanced greenhouse effect from gaseous emissions into the atmosphere, and El Niño events had been under investigation well before the 1997–1998 event (Trenberth and Hoar, 1996). The considerable oceanic warming associated with El Niño 97–98, with its record proportions, helped raise the possibility that global warming had had an influence on the occurrence and magnitude of El Niño 97–98. The question

FIGURE 4-8. This cartoon is an excellent indication of the public confusion created by the scientific debates over whether the strong El Niño 97–98 event was caused by global warming. (Reprinted with permission of Ed Stein, courtesy of the *Rocky Mountain News*)

of whether this was the case was raised throughout the 1997–1998 event, with widely different views expressed.

In September 1997, two scientific papers addressing the issue were published, and both questioned the existence of any connection between the recent ENSO conditions (pre-1997) and global warming, indicating that what had occurred was simply a part of natural climate variability (Latif et al., 1997; Rajagopalan et al., 1997). The positions of the authors of these papers differed from that of Trenberth and Hoar (1996).

At a summit meeting convened in California on October 14 to organize state mitigating activities to address expected El Nino problems, Vice President Al Gore proclaimed that the developing El Niño was enhanced by global warming. But, members of the press sanguinely noted that "Gore was trying to further the Clinton Administration's campaign to raise alarm bells over global warming" (Associated Press, October 14, 1997). Months later, as El Niño 97–98 faded from view in May 1998, another hypothesis was offered by Vice President Gore: that El Niño had accelerated the global warming trend (*Minneapolis Star Tribune*, June 8, 1998). Another view of the issue was that of an NCAR scientist who announced that the expected added rainfall resulting from El Niño would lead to the growth of more vegetation, and, in turn, the uptake of CO_2 by these added plants would lessen global warming (*Chicago Tribune*, October 31, 1997).

The debates continued into the winter. A reinsurance executive proclaimed that El Niño 97–98 was tied to global warming (*Cleveland Plain Dealer*, December 6, 1997), reflecting a belief, held in some insurance circles, that the recent extreme weather losses were a result of global warming (Changnon et al., 1996). A newly published scientific paper rebutting the two September papers stated that it was very unlikely that the ENSO events since the 1970s were due to natural variability and that they were likely a result of global warming (Trenberth and Hoar, 1997). Two days after this article appeared, scientists at the Lamont Doherty Laboratory announced that there was no relationship between global warming and El Niño's magnitude (*Indonesia Times*, December 8, 1997).

An NOAA scientist tied global warming and El Niño together as the explanation for record high global temperatures in 1997 (*Detroit News*, January 8, 1998). However, a group of NASA scientists challenged the correctness of the global temperature data used by NOAA, explaining that satellite-based global measurements showed a lowering of temperatures and that December 1997 was the coldest December on record since 1979 (Marshall Space Flight Center, January 20, 1998). An announcement by the World Meteorological Organization (WMO) late in January cited El Niño as the primary cause for the record high temperatures in 1997 and did not mention of global warming (*Des Moines Register*, January 31, 1998).

Numerous newspaper articles during February, March, and April 1998 continued to address both sides of the topic. Australia's federal science agency, CSIRO, announced that there was no connection between the two conditions

(ENN, March 12, 1998). By contrast, a university scientist studying El Niño and hurricanes concluded that there was likely a connection between global warming and the oddities associated with El Niño 97–98 (ENN, March 30, 1998).

The U.S. winter temperature for 1997–1998 was the second warmest in the past 103 years (Figure 4-9), helping to fuel the fires of speculation as to the cause. Early in April the head of NOAA stated, "You can't say that global warming was the cause of the strong El Niño," but in the same release the director of NOAA's climate data center said the effects of El Niño were amplified by global warming (*USA Today*, April 7, 1998). NOAA Administrator Baker also called the record warm winter of 1997–1998 a "window on the future if the global warming projections were correct" (*Minneapolis Star Tribune*, April 8, 1998), and NOAA officials stated that the model-based indications were that global warming would cause more weather extremes, including enhanced El Niños. The record high U.S. temperatures for January–May 1998 brought forth the question of whether some or all of the increase was due to El Niño's influence (Henson, 1999), and the proximity in time of the two strongest El Niños (1982–1983 and 1997–1998) was also seen as proof that human-caused climate change was the cause (Kerr, 1999). Some stated, as evidence of a global warming effect, that El Niño 97–98 could be making only a slight contribution to the record high temperatures in 1997–1998 because global temperature anomalies accompanying all past El Niño events had been much less dramatic than those in 1997–1998. The debate was well covered by the nation's press (Figure 4-10).

As El Niño 97–98 faded, the WMO announced in June 1998, "There is as yet no firm evidence of a connection between the frequency and intensity of El Niño events and global warming" (Associated Press, June 4, 1998). At the same time, Vice President Gore announced on June 8, 1998, "We set temperature records

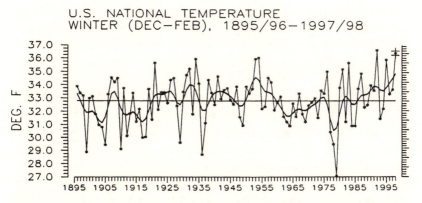

FIGURE 4-9. A graph depicting the 103-year history of winter (December–February) mean temperatures for the United States. This shows that the latest value, that for the El Niño winter of 1997–1998, was the second highest on record. (National Climatic Data Center)

Global warming not to blame for El Niño
Thursday, March 12, 1998

The much talked about rise in global temperatures over the past 100 years due to human-induced emissions of greenhouse gases is an unlikely cause of the spate of El Niño events during the 1990s, according to a scientist from Australia's federal science agency, CSIRO

WORLD **mate** NEWS

EL NIÑO AND CLIMATE CHANGE

A major climate event, the severe El Niño (see map above), and a major climate-related event, the establishment of the climate change, both

Los Angeles Times
APRIL 19, 1998

Connection between global warming and El Nino still murky

ENN
Environmental News Network

Beware El Niño Combined With Global Warming? Nah!

CLIMATE ALERT

Top 10 El Nino Events of this Century
Global Surface Mean Temperature Anomalies

Land

| | | | | 1.4 |
| | | | | 1.1 |
| | | | | 0.7 | Degrees F
| | | | | 0.4 |
| | | | | 0.0 |
1900 1920 1940 1960 1980 2000 -0.4

Volume 11, Numbers 1 and 2

USA TODAY
05/01/98- Updated

Is There a Relationship Between El Niño and Climate Change?

For the last 18 months or more, El Niño has been blamed for torrential rains in California, a crippling ice storm in Canada, huge forest fires in the Amazon, Kenya's most destructive floods in recent history, a prolonged power outage in New Zealand, floods and mudslides in Peru, a record dry spell in Hawaii, thousands of deaths from flooding and rain-related diseases in Central and East Africa

El Niño could be window on future

"Our computer modeling tells us that global warming may first manifest itself in changes in weather patterns," says D. James Baker, the head of the National Oceanic and Atmospheric Administration (NOAA). "In other words, this winter's El Niño is a taste of what we might expect if the Earth warms. It is a window on the future."

Gore connects El Nino to global warming at summit

SANTA MONICA, Calif. (October 14, 1997 5:34 p.m. EDT http://www.nando.net) -- Raising the pros killer floods and devastating droughts, Vice President Al Gore suggested Tuesday that global w may be adding muscle to El Nino weather effects.

FIGURE 4-10.   The scientific and policy debates over the possible relationship between the strong 1997–1998 El Niño event and human-induced global warming is illustrated by these clippings from the print media and scientific reports.

every month since January [i.e., February–May], and it appears that this general warming trend is making the effects of El Niño worse. This is a reminder once again that global warming is real and that unless we act we can expect more extreme weather in the years ahead" (*Minneapolis Star Tribune*, June 8, 1998). A White House official said Gore was using the situation to send a message to Congress that it was urgent to enact a $6.3 billion program of financial incentives and research aimed at cutting emissions of greenhouse gases.

In sum, the issue of the link, if any, between global warming and El Niño 97–98 remained the subject of continuing scientific debate. Sadly, the issue became a part of the major scientific-policy debate over the reality of greenhouse-induced global warming. Ironically, the El Niño–global warming debate may have added to the mounting public confusion about the reality of global warming and its causes. Chapter 3 explores the hypothesis that the El Niño 97–98 and its impact on the American public could be a signal event leading to increased belief and concern over global warming.

Differences in Predicted Impacts and Reactions of the Public and Decision Makers

Varying Impacts Predicted   Predictions of the physical, societal, and economic impacts expected to result from the El Niño–caused weather conditions began to appear soon after the initial seasonal weather predictions were issued in July–August 1997 (see Table 4-5). Early predictions of possible impacts focused on physical effects, such as more floods, altered water resources, and changes in other environmental conditions. During August and September 1997, FEMA issued several early warnings about expected severe flooding (FEMA, September 5, 12, 29, 1997).

Within weeks, these physical effects were being translated into potential economic impacts and societal problems, announced by a wide variety of persons, including economists, investors, brokers, and corporate salespersons. The early speculations ran the gamut from gloom and doom to "buy my product to make money as a result of El Niño's effects." For example, investment firms, seeking to reap benefits, advertised that El Niño would bring global food disasters and that there existed an opportunity to invest in crop futures and realize large profits (see chapter 6). Other advertisements focused on the prediction of a bad winter and hence the need to buy new snow tires, an odd choice since the predictions for the winter in most cold and snowy parts of the United States were for a mild winter with little snow.

The impact-oriented predictions emanating in the fall of 1997 offered very different and hence confusing views of the future effects of El Niño–generated weather conditions. Commodity experts predicted huge future demands for wheat (because of the droughts expected in Australia and China) and for coffee and cocoa (either droughts or too much rain in Central and South America), and they further predicted that the prices paid for corn and soybeans would skyrocket (*Cleveland Plain Dealer*, September 18, 1997). A well-known economist predicted that El Niño's effects would drive inflation and food prices "way up" and cause market instabilities (*Detroit News*, September 22, 1997). None of these predictions came true in 1997–1998.

A Chicago broker said the predicted heavy rains in California would increase certain food prices and the mild forecast for the Midwest would lessen natural gas demands (*Chicago Tribune*, September 25, 1997), both correct predictions. A private–sector meteorologist also correctly foresaw problems for the U.S. cotton crop in parts of the South (*Chicago Tribune*, September 25). Yet another private–sector meteorologist forecast abundant snowfall for the Rockies and a great ski season (*Cleveland Plain Dealer*, October 26, 1997), and a securities expert predicted natural gas prices would fall sharply (*Des Moines Register*, October 26, 1997).

An amazing variety of predictions of impacts continued to be issued as fall 1997 wore on. For example, one October news story, citing various experts, predicted salmon problems in the Northwest, rising sales for home repair businesses, flooding in Arizona and New Mexico, a bad vacation season for California resorts, cleaner air in California, and increased profits for commodity traders from investors who feared El Niño weather effects (*Minneapolis Star Tribune*, October 27, 1997). Vacation-related forecasts of all types continued to be issued. A November release indicated that vacationing in Florida would be wonderful, but actually that state experienced near-record winter rainfall, as shown in Figure 1-3 (*Des Moines Register*, November 16, 1997). In November, the *Wall Street Journal* (November 6, 1997) noted that investors and certain energy traders were betting on a cold winter. The next day an article predicted that gas companies would buy more natural gas and that insurance companies were headed for major losses as a result of El Niño storms in the Midwest (*Minneapolis Star Tribune*, November 17, 1997). This listing contains only a fraction of the public predictions of El Niño's effects. Many forecasts issued were not in agreement, a few turned out to be correct, and many were incorrect.

Another area of major predictive differences was the weather for the 1998 crop-growing season (see Figure 6-13). Pronouncements about the growing-season weather conditions and their effects on crop yields and prices had begun to appear in November (*Prairie Farmer*, November 1997). An Illinois climatologist predicted in January that high corn yields would occur in 1998 (Angel, 1998), and an Iowa climatologist also foresaw in February above-normal corn yields (Taylor, 1998). Many predictions with conflicting outcomes were issued by a variety of scientists and "agricultural experts" (*Farm Week*, March 9, March 16, 1998). USDA's chief economist predicted record soybean and corn yields for the United States, whereas a private-sector meteorologist predicted major droughts for the summer of 1998 and low crop yields (*Chicago Tribune*, February 23, 1998). Some of this variability reflected differences among various climate model projections; these had begun to be issued in midwinter, forecasting weather conditions for the summer and beyond.

Reactions to the Predictions   The reactions and responses to the seasonal weather predictions and the related forecasts of effects varied regionally. In California, which had suffered greatly from the 1982–1983 El Niño storms, many took the 1997 forecasts seriously and began planning in August and acting to get prepared. Meetings were held, many homeowners hired roofers to reinforce their roofs and the state and FEMA launched an active statewide mitigation program.

FEMA leaders reacted quickly to the ominous weather forecasts being issued by NOAA in June. A FEMA news release early in August pointed to the similarity between the new El Niño's strength and that of the 1982–1983 event and then described all the global damages created during 1982–1983 (FEMA, August 12, 1997). Three weeks later, a second FEMA news release described the 1997–1998 weather predictions from the CPC and related these to the conditions in 1982–1983 that caused U.S. losses of $2 billion and 160 lives (FEMA, September 5, 1997). Clearly, the agency saw NOAA's long-range predictions of heavy rains and storminess in parts of the nation as a major opportunity to launch preparatory mitigating actions and used scare tactics based on past losses to draw attention to the issue. FEMA's associate director testified in Congress early in September about how his agency was planning to work with states and local communities to mitigate damages from El Niño (Armstrong, 1997). The next day, another FEMA news release appeared, urging U.S. residents to prepare for the flooding expected to result from El Niño and pointing to the damaging floods of 1982–1983 in California, Utah, and Louisiana (FEMA, September 12, 1997). FEMA also announced it would conduct two "summits" for local/state officials, one in the West and one in the Southeast, in October, to launch major mitigative actions.

Further warning releases followed. A FEMA news release identified two recent hurricanes, Linda and Nora (which struck Mexico) as results of El Niño, revisited the large losses of 1982–1983, and announced that mitigative actions had begun in California (FEMA, September 29, 1997). FEMA then established on the Internet a "FEMA Loss Reduction Center" to serve as a national source for information on El Niño. This was followed by the California summit held on October 14 and attended by Senator Barbara Boxer, of California, Vice President Gore, and FEMA Director Witt; this event garnered national attention. The FEMA publicity endeavors may have been a major factor in creating the huge media attention to El Niño that became apparent during August and September 1997. Chapter 7 explores how this use of the climate predictions produced various activities and benefits in California.

In contrast to leaders in California, many decision makers in weather-sensitive sectors and the public elsewhere in the nation reacted indifferently to the predicted weather impacts. Chapters 5 and 6 address this issue and identify the reasons why many decision makers did not respond to the long-range climate forecasts. As shown in chapter 5, most of those who did react to the predictions realized major financial benefits.

A sampling of attitudes of 100 Iowans taken in November indicated that the majority doubted the El Niño–based weather predictions and would do nothing to prepare; they expected a normal winter (*Des Moines Register*, November 3, 1997). Several natural gas companies in the Midwest refused to react to the forecasts for a mild winter, saying, "You can not count on such a forecast, and we can not take a chance" (*Cleveland Plain Dealer*, November 9, 1997). Interviews with several persons charged with highway maintenance in the Midwest revealed that they were ignoring the regional forecast for mild conditions and unusually small amounts of snow. They bought normal amounts of salt because they felt they could not afford to be caught short, and salt producers indicated sales were good (*Cincinnati Inquirer*, November 9, 1997).

Another reaction to the impact predictions, and the fact that early storms (hurricanes and winter storms) seemed to support them, was a nationwide tendency to blame everything bad on El Niño, as explained in chapter 3. An early perception was that everyone in America needs a scapegoat for troubles: "We used to blame the Soviets, now we can blame El Niño" (*Chicago Tribune*, October 2, 1997). This same news writer correctly predicted that "everything that happens in the next ten months will be blamed on El Niño." The public's negative reactions to the El Niño impacts, along with a prediction for an additional possible but highly unlikely disaster in 1998, are well illustrated in Figure 4-11.

Another insightful writer noted how all types of personal problems—poor test scores, backaches, and bad hair—were the result of El Niño, which was labeled "the mother of all excuses" (*Minneapolis Star Tribune*, October 22, 1997). An article reviewing the impacts of El Niño's weather conditions, as the effect waned late in April, noted the hype and the ridiculousness of blaming everything on El Niño's influence (Crossen, 1998). This article also noted that major companies, including Toro (snow blowers), Disney (theme parks), Calaway Golf Company (golf clubs), and Dean Foods (fresh produce), had cited El Niño's influence on the weather as the reason for their sales troubles (Crossen, 1998). The story further noted that there had been major economic winners from the weather conditions, and it labeled El Niño as "a meteorological Robin Hood" because it provided winners in the northern sections of the nation while causing losers in the South.

Glantz (March, 1998) assessed the enormity of the El Niño hype, suggesting it was partly due to the news media and partly the result of scientists who oversold their predictions of damaging weather and many disasters. FEMA's efforts to win national support for mitigative efforts also helped create the impression that everything associated with El Niño 97–98 would be bad. Tactics tying the climate predictions of the 1997–1998 event to the conditions that prevailed during the 1982–1983 event, with its huge losses, were used by FEMA in a series of news releases issued during August and September 1997. Review of the

FIGURE 4-11.   During the peak of the media hype over El Niño came pronouncements, during March 1998, about the remote possibility that an asteroid might be approaching the earth. This cartoon captures the sense of the public's frustration over having to deal with the problems caused by El Niño, compounded by another possible natural disaster. (Permission to reprint given by Marshall Ramsey, Copley News Service)

numerous articles that addressed public reaction to El Niño suggests that the "blame El Niño for everything" habit began with stories by several news writers who reported that meteorologists and government agencies were blaming the early fall storms in and around the United States, as well as all the bad weather events everywhere else in the world, on El Niño.

El Niño Topics Addressed at Major Scientific Conferences

During the 1998–1999 winter, two major national scientific meetings were held, one by the American Geophysical Union (the AGU's fall meeting in December 1998), and one by the American Meteorological Society (the AMS's annual meeting in January 1999). Papers presented at these meetings had been chosen by their authors and submitted in mid-1998 and thus reflect, at the time El Niño 97–98 ended, the attention, ongoing research, and interests of the scientific community in El Niño 97–98 and ENSO in general. AGU membership includes atmospheric scientists, oceanographers, geologists, geographers, hydrologists, and some social scientists, whereas the AMS members are primarily atmospheric scientists, plus some hydrologists and oceanographers. Together, the members of these two societies represent broad and diverse scientific interests in major current topics, including global climate change, water and atmospheric pollution, and the El Niño phenomenon.

To assess the focus of ongoing and developing interest in and studies of El Niño 97–98, an analysis was made of the themes of the papers submitted to these two conferences (Figure 4-12). These papers reflect the main interests of the scientific community at the time, as El Niño 97–98 ended. The AMS meeting, which was a composite of seven scientific sector-oriented conferences held in parallel, included eighty-seven papers that pertained specifically to ENSO and/or El Niño 97–98, and the AGU meeting had 153 such papers. Both totals are significantly large and unmatched for any other comparable topic at either conference. In

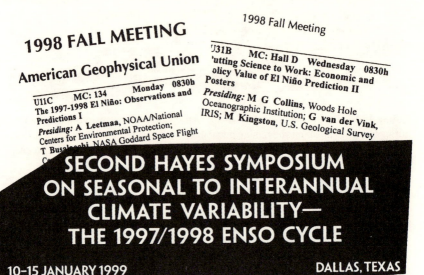

FIGURE 4-12. Examples of the materials distributed at major scientific conferences late in 1998 and early in 1999 that featured presentations of many papers about El Niño 97–98.

contrast, forty-four papers dealing with ENSO or some closely related physical feature were presented at the January 1997 annual meeting of the AMS, almost exactly half the number presented in 1998. This clearly illustrates a major surge of scientific interest relating to the occurrence of El Niño 97–98.

We analyzed the topics of each paper by title and abstract, and then classified the papers according to seven general themes. Table 4-7 shows the number of papers presented on each of these themes.

The two-meeting totals for these various themes reveal extensive scientific interest in four general areas: (1) defining/describing ENSO's physical characteristics, (2) long-term climate predictions based on El Niño conditions, (3) the effects of El Niño on the state of the atmosphere and on surface weather conditions, and (4) the impacts of El Niño–generated weather on physical systems. An additional question is, How well did these varied interests of the scientific community mesh with the five scientific issues that had emerged during El Niño 97–98?

The issue of attribution; that is, which weather events were caused by El Niño, was given considerable attention, as reflected in the large number of papers (sixty-six) that addressed the effects of El Niño on various weather conditions. Several papers assessed different procedures (models) for identifying which conditions (such as storms) were related to El Niño. Different views were presented about which events were totally or partly related to El Niño, reflecting the different views that were expressed by scientists during El Niño 97–98.

Concerns about the disparities in climate predictions based on El Niño that appeared during the event were only indirectly addressed in some of the thirty-five papers that focused on climate predictions based on El Niño. Several scientists from groups that had issued differing predictions during El Niño (particularly for the spring and summer 1998) gave talks at the meetings, but the subject of how their differences had affected users of forecasts and the general public was not addressed. There was however, debate over methodological differences in predictive techniques.

TABLE 4-7. Major themes of scientific papers on El Niño conditions presented at the 1997 Fall Meeting of the American Geophysical Union and the January 1998 Annual Meeting of the American Meteorological Society.

Theme	AMS papers	AGU papers
Physical characteristics of ENSO and El Niño 97–98	19	7
Long-term climate predictions based on El Niño 97–98 (e.g., how done, accuracy)	13	22
Satellite sensing of oceanic-ENSO conditions	7	14
Uses of El Niño–based climate predictions	4	16
Effects of El Niño/ENSO conditions on weather and atmospheric conditions	28	38
Impacts of EL Niño weather on physical systems (land forms, hydrology, and crops)	13	52
El Niño 97–98 and global-warming	3	4

Some attention was paid at both meetings to rating the magnitude of El Niño 97–98. Some of the twenty-six papers assessing the "physical characteristics of El Niño 97–98" specifically presented dimensional data and compared the 1997–1998 conditions with those of earlier El Niño events, but there was no debate as had occurred during the evolution of the 1997–1998 event. Most papers presented at the 1997 AMS meeting had dealt with the physical features of ENSO events.

The question of whether global warming had influenced El Niño 97–98 was addressed by most of the seven papers on this theme. Both sides of the debate were expressed, but, interestingly, the topic was not extensively addressed at either conference.

The issue of different predicted impacts and public/user reactions to El Niño received little attention at the two meetings. Twenty papers assessed various uses of El Niño–based predictions, but meager attention was given to the wide differences expressed during El Niño 97–98 about the potential physical and social outcomes. This lack of attention is likely a result of the fact that many of those who had issued impact predictions were either not the type of scientists who attended these two meetings or were nonscientists; many who issued impact predictions were economists, businessmen, government bureaucrats, or private-sector meteorologists.

Several papers at the two meetings addressed two themes not identified as key issues during El Niño. First, there were sixty-five papers about the actual measured geologic (land surface) and hydrologic impacts (e.g., floods, streamflow) of the weather attributed to El Niño 97–98. These impacts were not a contentious scientific issue. There were also twenty-one papers on the use of satellite sensors to measure sea surface temperatures and other ENSO conditions; this also was not a controversial topic.

SUMMARY

Without question, El Niño 97–98 became the most discussed climate event in the nation's history. Everyone had opinions about El Niño, and it became a household phrase, used in either a serious or a humorous way to explain problems or failures of any type. It created a new national interest in weather and climate and their causes, as well as in the use of long-range predictions. By and large, the event and its successful climate predictions created a new and better public image for the atmospheric and oceanic sciences.

The major findings from this scientific assessment were sorted into three categories: (1) those related to presentation of scientific information including the source and how transmitted, (2) those pertaining to the climate and impact predictions, and (3) those that addressed the major scientific issues that arose during El Niño.

Scientific Information

More than thirty government, university, and private institutions, and often scientists and administrators working for them, issued some form of predictions of climate conditions and/or impacts or other forms of scientific information about El Niño. Having so many sources of predictions created great potential for contradictory forecasts, and this is in fact what occurred, leading to confusion among the public and among decision makers.

The volume of scientific information issued by the news writers and by topical journals was sizable. Newspaper articles, at first written by science writers, were, by fall 1997, being authored by feature writers (as public acceptance of the El Niño phenomenon became essentially complete). The "blame everything on El Niño" theme appeared in September and October and was driven by information presented in the newspapers that in turn was probably derived from a series of news releases from NOAA and FEMA. Several FEMA releases repeatedly used scare tactics and references to large losses in the past to get public attention and to drum up interest in pursuing mitigative actions.

Producers and users of scientific information realized that the Internet allowed anyone anywhere to present scientific information without having any qualifications. Much of the material about El Niño offered by several scientists and institutions on the Internet was repetitious and redundant. Included were definitions of El Niño, as well as seasonal climate predictions based on El Niño conditions. A very large variety of sources of information existed on the Internet, and some of these had questionable scientific credentials.

Some of NOAA's releases created the potential for confusion because many NOAA entities and staff were presenting information at the same time. Some of the information overlapped as it pertained to forecasts, potential impacts, and possible causes of El Niño. Some statements issued by NOAA staff disagreed with those issued by other NOAA staff. For example, some releases and statements carried contradictory views on controversial issues such as global warming and its influence on El Niño, and several offered contrasting views of El Niño's influence on the atmosphere and its role in producing adverse weather events like tornadoes.

By January–February 1998, many news releases with a "science" content and containing climate and/or impacts predictions were being written by journalists who mentioned no scientists as sources for their reports. A common theme in the media about the impacts of El Niño was that the outcomes were all "bad," and reporters evinced surprise when a "good" impact was found. News writers, without acknowledging scientific sources, made varied interpretations of the predictions, their outcomes, and their verification, and this type of reporting developed and became common during winter 1997–1998. In spring 1998, news writers reverted to the use of scientists for input to news stories, a situation that was linked to the emergence of differences in the scientific predictions being offered about the weather in the spring and summer to come.

Seasonal Climate and Impact Predictions

The climate models used to generate the long-range predictions of the onset and strength of El Niño did not predict these phenomena well. However, the models did predict the event's magnitude and its duration after it started developing in June 1997. The predictions about El Niño's effects on seasonal climate conditions began appearing in June 1997 and continued through spring1998. The predictions for the fall, winter, and early spring conditions were amazingly accurate for many parts of the nation, but those for the growing season of 1998 varied widely depending on the climate model used, reflecting no great skill on the part of the predictors.

There was the potential for confusion in various user communities because there was an official climate forecast (although it is easy to miss the "official" labeling), as well as several other forecasts offered by scientific institutions that were issuing "experimental" or research-based forecasts. Many of these were issued without disclaimers. The nonscientist users of the predictive information could easily have missed the difference between experimental and official predictions. In midwinter 1997–1998, widely different predictions began to appear about when the current El Niño would end and about what conditions (normal or La Niña–driven) would follow. The uncertainty these different predictions created nullified for some users of long-range forecasts the credibility generated by the highly correct predictions for the fall and winter conditions.

Predicted environmental and societal impacts of El Niño conditions during the summer and early fall of 1997 initially focused on physical outcomes such as flooding. But by midfall, predicted impacts ran the gamut from disaster for many business sectors to major economic benefits. The forecasts included detrimental effects on agriculture, the citizenry (financially and health-wise), the energy industry, recreation and vacation areas, commodity markets, and the nation's overall financial well being as a result of excessive rain and storms.

Scientific Issues

The major scientific disagreements over El Niño 97–98 centered around four issues. The first concerned the differences in the oceanic and climate conditions predicted for the spring, summer, and end of 1998. This disagreement was a contrast to the close agreement that characterized predictions issued in 1997 for the fall and winter 1997 conditions.

The second major disagreement was among scientists and centered around which weather conditions and storm events of the October 1997–May 1998 period were totally, partially, or not at all the result of El Niño's influence on the atmosphere. Debates and controversies were particularly common in the media's presentations of explanations for the causes of many tornadoes. This situation was compounded by the media's tendency to blame everything that happened with the weather on El Niño.

A third confusing issue centered around whether El Niño 97–98 and its record size were tied to global warming. Widely different views came from scientists, administrators, and the White House. The debate became engulfed in the larger national scientific-economic-political controversy over global warming and what to do about it.

The fourth area of major scientific disagreement centered around the widely different impacts foreseen as part of the varying climate predictions being issued. This especially affected predictions of conditions for the 1998 growing season and the possible ramifications for agricultural production that year. The debate resulted from the wide differences in the model predictions, which suggested a continuing El Niño, a return to normal condition, or a sudden development of La Niña conditions in 1998. However, even when the predictions being issued for the 1997–1998 winter conditions were all in agreement, widely different impact predictions were issued by a wide variety of sources, a reflection of the general lack of understanding of how the weather affects the nation's complex societal and economic systems.

APPENDIX

The institutions and individuals issuing predictions of climate conditions in the U.S. and/or impacts from predicted climate conditions, based on El Niño 97–98.

Foreign countries and international organizations
    World Meterological Organization (WMO)
    Argentine Weather Bureau
    Australia Bureau of Meterology
    Environment Canada
    Thailand Meterological Department
    University of British Columbia, Canada
    University of Oxford, United Kingdom

Federal agencies
    Department of Commerce/National Oceanic and Atmospheric Administration
        Climate Prediction Center of National Weather Service
        Pacific Marine Experimental Laboratory (PMEL) and its TAO Group
        Secretary of Commerce and NOAA Administrator
        Office of Global Programs
        Climate Diagnostics Center at CIRES

NWS Field Offices across the nation
The Climate Modeling Branch of NCEP/NWS
Severe Storm Forecast Center
U.S. Geological Survey
Federal Emergency Management Agency
National Science Foundation
U.S. Army Corps of Engineers
National Aeronautics and Space Administration
U.S. Department of Agriculture

State and regional agencies or institutes
Western Regional Climate Center
Midwestern Climate Center
Southeastern Regional Climate Center
State climatologists in Illinois, Iowa, Virginia, and Michigan
State agencies—California, Florida, Texas, and Illinois

University groups/centers/individuals
Universityof Alaska's Natural Heritage program
University of California at Irvine
Columbia University's Lamont Doherty Earth Observatory
Florida State University's Center for Ocean-Atmospheric Prediction (COAP)
  Studies
San Francisco State University
Georgia Institute of Technology
Ohio State University
Southern Illinois University
Mississippi State University
Michigan State University
Colorado State University
University of Michigan
Iowa State University
Creighton University
University of California at Los Angeles
University of Illinois
George Mason University's Center for Ocean-Land-Atmosphere Studies, El Niño
  Resources Center

Private sources (media, business, or research)
CNN News, AP News, UPI
Major newspapers across the nation
Private weather firms

National laboratories and comparable independent research centers
Scripps Institute of Oceanography, Experimental Climate Prediction Center
National Center for Atmospheric Research (NCAR)
Goddard Space Center
Jet Propulsion Laboratory at California Institute of Technology
International Research Institute for Climate Prediction

REFERENCES

Angel, J. January 31, 1998. Climatologist sees above trend yields. *Champaign-Urbana News Gazette*, p. 1.

Armstrong, J. September 11, 1997. *Statement of Associate Director for Mitigation, FEMA*. House Subcommittee on Energy and Environment, FEMA News Release, Washington, DC.

Associated Press. October 14, 1997. Gore connects global warming at summit.

———. December 7 1997. El Niño fueled storm socks California.

———. January 11, 1998. Canadians reel from icy disaster.

———. June 4, 1998. Pesky El Niño is singing its swan song, forecasters say.

Barnston, A. G., Glantz, M. H., and He, Y. 1999. Predictive skill of statistical and dynamical climate models in SST forecasts during the 1997–98 El Niño episode and the 1998 La Niña onset. *Bulletin Amer. Meteoro. Soc.*, 80, 217–243.

Champaign-Urbana News Gazette. March 1, 1998. El Niño's wrath, p. A7.

———. March 15, 1998. El Niño's wrath, p. A11.

———. June 7, 1998. Official: Mexico's fires most serious yet seen, p. A6.

Changnon, S., Changnon, D., Fosse, E., Hoganson, N., and Roth R., Sr. 1996. *Impacts and Responses of the Insurance Industry to Recent Weather Extremes*. Mahomet, IL: Changnon Climatologist.

Chicago Tribune. September 14, 1997. Hurricane weakens off Mexico, aims for California, p. 9.

———. September 18, 1997. Threat of El Niño brings boom season to California roofers, p. 3b.

———. September 25, 1997. El Niño already causing headaches; it is worrying traders, p. 1.

———. October 2, 1997. Blame it on El Niño, p. 17.

———. October 14, 1997. California El Niño summit shares ideas on weathering the storm, p. 2.

———. October 31, 1997. El Niño may help slow global warming, p. 13.

———. February 23, 1998. USDA forecasts bumper corn and soybean crops, meat and poultry glut, p. 3.

———. February 23, 1998. El Niño linked tornadoes kill dozens in Florida, p. 1.

———. April 13, 1998. El Niño has twister experts guessing, p. 5.

Cleveland Plain Dealer. September 18, 1997. El Niño set to boost grain prices, p. 1C.

———. October 26, 1997. Why El Niño warms hearts on ski trails, p. 5K.

———. November 9, 1997. Amid mild winter forecasts, utilities prepare for the worst, p. 6B.

———. December 6, 1997. Insurers worry, p. 10B.

———. February 2, 1998. Nothing average about El Niño effect, p. 3.

———. March 1, 1998. When winter went AWOL—set records for least snowfall and warmest temperatures, p. 1A.

Cincinnati Inquirer. October 12, 1997. Giant El Niño rocking science and financial worlds—planners brace for the worst case, p. D01.

———. November 9, 1997. Tristate road crews aren't lulled—salt supplies disregarding El Niño, p. B03.

Climate Prediction Center. August 4, 1997. *Special Climate Summary*. Washington, DC: CPC.

———. February 12, 1998. *Prognostic Discussions for Long-Lead Outlooks*. Washington, DC: CPC.

CNN. May 26, 1998. El Niño backing off, NASA pictures reveal (Internet).

COLA. December 1997. *Forecasts from December 1997–El Niño Gives Way to La Nina*.

Crossen, C. April 17, 1998. And exactly what isn't El Niño's fault? *Wall Street Journal*, p. 1.

Des Moines Register. September 16, 1997. Data show El Niño is beefing up, p. 10.

———. October 26, 1997. Natural gas stocks may be only thing El Niño won't affect, p. 1.

———. November 3, 1997. Do you think Des Moines will have a bad winter because of El Niño? p. 3.

———. November 16, 1997. Will El Niño be a fair or foul weather friend? p. 1.

———. January 31, 1998. El Niño's mischief expected. p. 1.

———. April 6, 1998. Will El Niño lessen the chance for tornadoes? p. 3.

Detroit News. September 22, 1997. Investing: Markets reacting to any old noise, p. F9.

———. January 8, 1998. 1997 part of global warming trend, p. A5.

———. March 20, 1998. Farewell to the winter that wasn't, p. E1.

———. May 7, 1998. El Niño not to blame for all weird weather, p. A15.

Douglas, A. March 7, 1998. Weather uncertainties loom over 1998 planting. *Farm Week*, p. 5.

Douglas, P. September 2, 1997. *Minnesota Star Tribune*, El Niño Weather p. 6b.

Environmental News Network. March 12, 1998. Global warming not to blame for El Niño (Internet).

———. March 30, 1998. Current El Niño called a strange one. (Internet).

Farm Week. March 9, 1998. Weather uncertainties loom over 1998 planting, p. 5.

———. March 16, 1998. Hot year for 1998 crops. p. 6.

Federal Emergency Management Agency. August 12, 1997. *Strong El Niño could disrupt U. S. winter weather patterns*. Washington, DC: FEMA.

———. September 5, 1997. *El Niño of 1997–98 could resemble the destructive 1982–83 event*. Washington, DC: FEMA

———. September 12, 1997. *U. S. residents urged to prepare in advance for potentially heavy rains and flooding expected to accompany this year's powerful El Niño*. Washington, DC: FEMA.

———. September 29, 1997. *El Niño preparedness appeal*. Washington, DC: FEMA.

———. October 13, 1997. *The making of El Niño summit*. Washington, DC: FEMA.

Glantz, M. R. March 1998. El Niño forecasts: Hype or hope. *Network Newsletter*, p. 1.

———. August 1998. *Network Newsletter* (Editorial), p. 1.

Gray, W. November 29, 1997. Hurricane season oddly quiet in 1997. *Champaign-Urbana News Gazette*, p. 5.

Grenci, L. December 1997. A month of tyranny. *Weatherwise*, 50, 44.

Henson, R. 1999. Hot, hotter, hottest: 1998 raised the bar for global temperature leaps. *Weatherwise*, 52, 34–37.

Indonesia Times. December 8, 1997. Connection between global warming and El Niño murky, p. 3.

International News. March 10, 1998. El Niño increases storm frequency and intensity, p. 2.

Kerr, R. A. 1999. Big El Niño's ride the back of slower climate change. *Science*, 285, 1108–1109.

Latif, M., Klaman, R., and Eckert, C. 1997. Greenhouse warming, decadal variability or El Niño? An attempt to understand the anomalous 1990s. *J. Climate*, 10, 2221–2239.

Le Comte, D. 1999. A warm, wet, and stormy year. *Weatherwise*, 52, 19–28.

Marshall Space Center. January 20, 1998. *December 1997 is coldest month on record in the stratosphere.* Washington, DC: Author.

McPhaden, M. J. 1999. Genesis and evaluation of the 1997–98 El Niño. *Science*, 283, 950–954.

Minneapolis Star Tribune. October 22, 1997. Blame El Niño, p. 8B.

———. October 27, 1997. Taking El Niño to the bank, p. 5A.

———. November 17, 1997. There's no shortage of misconceptions associated with El Niño, p. 1D.

———. February 27, 1998. El Niño disrupts the weather and confounds scientists: The debate over effects of El Niño rages unabated, p. 21a.

———. June 8, 1998. El Niño accelerates global warming trend, p. 3A.

NBC. February 15, 1998. What's ahead for El Niño? (Internet).

Nature. July 10, 1997. El Niño forecast fails to convince skeptics, 385, 108.

Powers, K. March 31, 1998. Rare tornadoes strike Minnesota. *Cleveland Plain Dealer*, p. 7A.

Prairie Farmer. November 1997. Marketing moves, p. 5.

Price, N. October 9, 1998. Powerful Pauline barrels into Mexico: El Niño possible cause of hurricane's wrath. *Cincinnati Inquirer*, p. 1.

Rajagopalan, B., Lall, U., and Cane, M. 1997. Anomalous ENSO occurrences: An alternate view. *J. Climate*, 10, 2351–2357.

St. Louis Post-Dispatch. September 4, 1997. El Niño alert: Professor turned weather man forecasts cold, stormy winter, p. 1A.

———. November 22, 1997. El Niño hits home, p. S2.

San Jose Mercury News. March 16, 1998. Expect more from El Niño, p. 1.

Southern Illinoisan. October 26, 1997. El Niño might not affect us, but then again, p. 4A.

Taylor, E. February 2, 1998. Climatologist sees above trend yields. *Farm Week*, p. 5.

Trenberth, K. E. 1999. Development and forecasts of the 1997/98 El Niño: Clivar scientific issues. *Exchanges*, 3, 4–13.

Trenberth, K. E., and Hoar, T. 1996. The 1990–1995 El Niño-Southern Oscillation event: Longest on record. *Geophysical Res. Letters*, 23, 57–60.

———. 1997. El Niño and climate change. *Geophysical Res. Letters*, 24, 3057–3060.

University of Illinois. November 4, 1997. *El Niño does not increase risk of large storms on the Great Lakes.* Champaign, IL: University of Illinois.

University of Michigan. October 24, 1997. *UM researchers find links between El Niño and weather in Great Lakes region.* Ann Arbor, MI: University of Michigan.

UPI. March 9, 1998. Can't blame it on El Niño. (Internet).

USA Today. October 28, 1997. El Niño gave blizzard much of its strength, p. 4A.

———. December 22, 1997. Texas braces for second round of rain and snow, p. 2.

———. January 8, 1998. Florida having a wet dry season, p. 8.

———. February 3, 1998. Storms batter California and Florida, p. 1.

———. February 19, 1998. El Niño will continue into summer, p. 6.

———. February 28, 1998. By shifting jet stream, El Niño worsens storms, p. 3.

———. March 25, 1998. El Niño's winter no more costly than others, p. 1.

———. April 3, 1998. El Niño unlikely to affect twister paths, p. 1.

———. April 7, 1998. El Niño could be window on future, p. 1.

———. June 26, 1998. Florida fires worse than hurricane, p. 1–2.

U. S. Water News. August 1997. Is another aberrant El Niño weather event on the way? p. 14.

———. January 1998. El Niño does a favor and dismantles Atlantic hurricane season, p. 5.

Wall Street Journal. November 6, 1997. Cold weather on the way? Some bet on it, p. C1.

Washington Post. February 24, 1998. El Niño was a major factor in tornadoes, p. A14.

———. February 26, 1998. Is global warming linked to strength of latest El Niño? p. 3.

5

 ## Who Used and Benefited from the El Niño Forecasts?

DAVID CHANGNON

he long-range seasonal climate forecasts based on El Niño 97–98 conditions and issued from June through August 1997 for the fall, winter, and early spring conditions across the United States were accurate for many parts of the nation (see chapter 2). An important question concerns whether decision makers in weather-sensitive public and private organizations used these El Niño–derived seasonal forecasts. Most seasonal forecasters viewed with great confidence the predictions of a strong El Niño and associated precipitation, temperature, and storm anomalies expected across the United States. From their perspective, it was an opportune time to use and, presumably, to benefit from the forecasts.

Our assessment of a large group of potential users of the seasonal forecasts sought to identify who used and did not use the forecasts, the reasons for their use or nonuse, and the applications and potential value of the forecasts derived from their use. Sector differences were assessed by sampling decision makers in agribusiness, water resources, utilities, and other sectors. Results of such use and nonuse investigations will help develop better, more effective strategies for disseminating climate forecasts (Pfaff et al., 1999). Another objective of this study was to understand the perceptions decision makers had of seasonal forecasts and how the successful predictions based on El Niño 97–98 may have modified those perceptions. Figure 5-1 presents a typical humorous media view of the forecasts.

A survey of individuals was conducted to gather the desired information about how the seasonal forecasts based on El Niño 97–98 were obtained, evaluated, and incorporated into decisions. The study was designed to focus on decision makers in weather-sensitive positions and to employ sampling techniques tested and developed in prior surveys. These previous studies had developed, tested, and used questionnaires as the tool by which to gather information about the use of

"Board up the houses! Round up the stock! Lock up the women and children! Here comes El Niño!!"

FIGURE 5-1. This cartoon, published in August 1997, illustrates the rapidly developing wide interest and concern about how to react to the El Niño forecasts that were issued in the summer of 1997. (Reprinted with permission of the *Sacramento Bee*)

climate information by weather-sensitive users in water resources, agribusiness, and utilities (Changnon, 1982, 1991, 1992; Changnon and Changnon, 1990; Changnon et al., 1988, 1995; Sonka et al., 1992). These past assessments found that 65 percent or more of the sampled decision makers were not using available climate forecasts and that there were several reasons for this, including an inability to incorporate uncertain information (inherent in climate forecasts) in decision models (Brown et al., 1986; Brown and Murphy, 1987; Glantz, 1987). In all these assessments, a key reason for not using the seasonal forecasts was the lack of models that related climate conditions and forecasts quantitatively to important physical conditions (e.g., soil moisture, runoff in a basin, crop yields across a region) and to critical economic outcomes (e.g., sales, retention of seed surpluses, construction plans, advertising schedules, irrigation scheduling, purchases). These studies also emphasized that many decision makers in weather-sensitive groups who utilized available climate information and forecasts required assistance from professionals who understood the information and its specific use and who could help interpret it for incorporation into the decision-making process (Vogelstein, 1998).

This chapter provides a profile of those chosen for interviews, assesses the results of the interviews, and summarizes the findings. The findings provide information useful to those actively involved with developing, formatting, and disseminating climate forecasts and related information.

THE SAMPLING PROCESS

Sampling was designed around use of face-to-face or telephone interviews. Most sampling was done using telephone interviews, because sampling of decision makers distributed across the United States was desired and available funds did not permit extensive travel for many face-to-face interviews.

A questionnaire was developed using previous examples (Changnon et al., 1995). A draft questionnaire was developed in April 1998 and was pretested using three individuals in weather-sensitive positions. Their comments were used to alter and develop the final questionnaire. This questionnaire led those being interviewed through a decision tree of questions and responses that could be completed in about thirty minutes. Past experience revealed that most decision makers did not have time for more time-consuming interviews.

The questionnaire had four sections. The first familiarized respondents with the project. The project was described, and a description of individuals' organizational responsibilities and the influence of weather on their jobs was defined. Second, the reasons why interviewees did or did not use the seasonal forecasts related to El Niño 97–98 were elicited. If the individual used the forecasts, questions in part three defined the forecast applications and their value, if known. If the forecasts were not used, the reasons for nonuse were defined. Finally, individuals being interviewed were given an opportunity to discuss any thoughts concerning the official long-lead forecasts and other weather/climate information they used to aid in their decisions.

The initial part of the questionnaire aimed at providing a short but informative background description of the project and its goals. Those conducting the interviews introduced themselves, briefly described the project (the stated objective was to obtain a "scientific assessment" of use/nonuse), and let the interviewees know how they had been identified as people who might have incorporated the El Niño 97–98 forecasts into their decisions. Potential interviewees were asked if they had time to answer the questions. If the potential interviewees agreed, the interviewers further described the goals of the project and how decisions makers might benefit from the results of the study. At that point, the interviewer reviewed key information (e.g., forecasts, regions in the United States expected to be impacted by climate anomalies) associated with El Niño 97–98. Interviewees were then asked to briefly describe their area of responsibility within their organization. They were also asked how weather influenced what they did, and in what ways.

Once this background material had been covered, the interview moved to a primary question: "Did you or others in your organization utilize the information related to the forecasted climate conditions associated with El Niño 97–98?" If the answer was yes, the next group of questions concerned (1) when the users became aware of the forecast, (2) where they obtained the information, (3) their initial attitude toward the long-range forecasts, and (4) how long it took before the user developed some level of "understanding" of the forecast. For those who used or even considered using the climate forecasts (but who ultimately did not use them), questions sought to define specifically how and when they used the climate forecast information. They were asked whether (1) it was difficult for their management to accept the use of the forecast; (2) there had been sufficient lead time to implement the forecast information in their decisions; (3) the information was used in operational decisions, short- or long-term planning, or in other actions; and (4) there had been beneficial outcomes (e.g., economically, environmentally) from using the forecast information and, if so, what specific types of benefits and monetary advantages were achieved.

If the interviewees indicated that they did not use the El Niño–related climate forecasts, questions assessed the reasons for nonuse. These included questions about whether the interviewee did not know about the forecasts in time to utilize them, did not trust the source of information, was concerned about the subpar levels of forecast accuracy, had difficulty in interpreting or implementing information into decisions, and feared that upper management would not allow use of an uncertain product in important decisions. Other reasons were also sought.

After completing these parallel series of use or nonuse–related questions, interviewees were asked whether the reasonably accurate long-range predictions based on El Niño 97–98 would cause them and/or their organizations to alter their use of climate forecasts in the future. Finally, the interviewees were asked if they had any additional comments or questions about long-lead climate forecasts and other weather-related tools. They were also asked if they could be contacted again if necessary to clarify any issues. At the end of the interviews, they were thanked for their assistance.

The study's goal was to interview between 75 and 90 decision makers who were (1) pursuing weather-sensitive tasks and (2) employed in agribusiness, water resources, power utilities, and other weather-sensitive sectors. We also sought to sample persons dispersed across the United States. The interviewing process began late in May 1998, as the El Niño influence on the weather conditions of the United States was diminishing, and the interviews were completed by October 1998. It was important to interview individuals as soon after the event as possible so that various details involved in their decisions would be fresh in their memories. Nearly 90 percent of the interviews were completed between July 1 and August 31. Seventy-three of the eighty-seven interviews were conducted by phone. The interviews were done by two persons experienced in climate-user assessments (Changnon, 1992; Changnon et al., 1988, 1995).

Decision makers interviewed were identified by several means. The scientists working on this project helped provide names of those who might have used climate forecast information related to El Niño 97–98. The help of persons interviewed in prior assessments of climate forecast users (Changnon, 1982, 1992; Changnon et al., 1995) was also sought; some of these persons provided names of others to interview. Other contacts were found through reports and newspaper articles about El Niño. Staff at the National Weather Service (NWS) and Service climatologists were able to provide various insights about use/nonuse of the El Niño forecast in their sectors, as well as potential names of persons in their regions to contact. Interestingly, many climatologists in state climate offices or regional climate centers were unable to provide names of decision makers who might have used the El Niño–related forecasts.

Through this process, we compiled a list of 137 potential individuals to interview. The majority of those interviewed were unknown to those doing the interviewing before the start of the sampling. Most of the 87 persons ultimately interviewed were chosen randomly. More than 125 individuals were asked, and 87 agreed to be interviewed. Generally, the interviews lasted between twenty-five and thirty minutes, but interviews with those who had a great interest in the event often lasted an hour.

Table 5-1 presents the number of individuals interviewed by sector (e.g., agribusiness, power utilities, water resources). Reasonably large samples were collected for the three sectors of primary interest—agribusiness, water management, and power.

Table 5-2 presents the states where the interviewees resided. Most of those interviewed were located in regions where some type of seasonal climate anomaly (such as above- or below-average precipitation or temperature) had been predicted (and ultimately occurred). Most climate regions of the United States were represented, with data collected from twenty-eight states.

USE AND NONUSE OF PREDICTIONS BY 87 DECISION MAKERS

The interviews revealed the complexity and limitations associated with the use and nonuse of the seasonal forecasts. Most of the eighty-seven persons interviewed were interested in discussing the El Niño phenomena, the 1997–1998 event and the forecasts based on it, and how the event had impacted their tasks. For the first time the El Niño–based forecasts provided most of the decision makers

TABLE 5-1. Number of persons interviewed in each sector.

Sector	No. interviewed
Agribusiness	17
Power utilities	22
Water resources	25
Climatologists advising select customers	11
Others (insurance, transportation, emergency preparedness)	12

we interviewed with an opportunity to develop and/or to implement operational or planning decisions months in advance of a specific season of concern.

Survey results revealed that how a forecast was used was not always a simple yes or no answer. The interviews produced a number of expected conclusions about El Niño and the related seasonal forecasts, including the fact that using an uncertain product within most organizations is often very complex. Overall, the enthusiasm over the accuracy of the forecasts exhibited by those interviewed indicated that interest in and use of climate forecasts would increase in the future.

The initial question asked of those interviewed was whether they had used the El Niño–related seasonal forecasts in making any of their organization's decisions. Some gave a yes or no answer to this question, but responses from several were a qualified yes-then-no response. That is, many interviewees had closely examined and carefully considered the forecasts for a period of time, but because of various concerns held by them or by their organization's management, in the end the forecasts were not integrated into actual operational or planning decisions. Thus, an unexpected theme relating to usage was that "the forecast was examined and carefully considered, but never actually used." Of the eighty-seven interviewed,

TABLE 5-2. Number of persons interviewed in each state.

State	No. interviewed	State	No. Interviewed
Arizona	1	Montana	7
California	13	Nebraska	2
Colorado	5	Nevada	2
Florida	7	New Mexico	1
Georgia	2	New York	1
Idaho	1	North Carolina	3
Illinois	11	Ohio	1
Indiana	1	Oklahoma	2
Iowa	7	South Carolina	1
Kansas	1	Texas	6
Kentucky	1	Utah	1
Louisiana	1	Virginia	1
Michigan	2	Washington	2
Missouri	2	Wisconsin	2

- forty-one (47 percent) used the seasonal forecasts in making an operational and/or planning decision
- twenty-nine (34 percent) carefully examined and seriously considered using the forecast but ultimately decided to not incorporate it in a decision
- seventeen (19 percent) said that they immediately decided not to use the forecasts to alter any of their weather-related decisions.

The seventy persons who said they used (yes with action) or examined (yes with no action) the long-range forecasts were asked to list the source(s) of the forecasts they assessed. Many relied on more than one source for the seasonal forecasts; hence, the values for the seven sources identified in Table 5-3 total more than seventy. Forty-four individuals said they acquired and examined forecasts from two or more sources, often with the objective of comparing forecasts.

Although sixty-one of the seventy yes respondents (87 percent) said that they used or considered using the Climate Prediction Center (CPC) forecasts, a good number (nineteen) also obtained El Niño–based forecasts from private meteorological firms. The CPC forecasts made available on December 19, 1997 are presented in Figure 5-2. Twenty-seven of the forty-four (61 percent) who obtained forecasts from two or more sources reported they found conflicting information, particularly after December 1997. Although the fall and winter forecasts for regional temperature and precipitation anomalies were generally consistent among most forecast suppliers (see chapter 4), the more specific information pertaining to the exact timing and magnitude or placement of the forecasted anomalies often varied between forecasts. The resulting uncertainty often led to the nonuse of the forecast information by those who had carefully considered their use. This points to the importance for forecasters in the government and at scientific institutions of presenting comparable seasonal forecast information. Perhaps the official CPC forecast should carry a statement about experimental research-based forecasts that find their way onto the Internet and into the media and contrast these to the official forecast.

TABLE 5-3. Sources of seasonal climate forecasts and number of acquisitions per source.

Source	No. of acquisitions
Climate Prediction Center	61
Other government agencies	10
Weather Channel/TV news	10
Newspaper stories	6
Academic sources/research institutions	11
Private meteorological firms	19
Others in parent organization	4

El Niño/Southern Oscillation (ENSO)

ENSO Advisory

ENSO Update
• Updated on Monday

MEDIA GRAPHICS

Special El Niño Summary:
Impacts and Outlook

El Niño impacts on
1997 Hurricane Season

Sea Surface Temperature
Forecasts

NCEP

Official NWS Forecast

Coupled
Ocean-Atmosphere Model
Forecast

OTHERS

Lamont-Doherty Equatorial
Pacific Forecast

CDC Linear Inverse
Modeling Forecast

Regional Analyses

Asia

Australia

South America

Mexico and Caribbean

South Africa

United States

HISTORICAL FEATURES

 UPDATED

Individual State El Niño

CURRENT STATUS
Updated December 19, 1997

Strong warm episode (El Niño) conditions have persisted in the tropical Pacific since June 1997. Sea surface temperatures throughout the equatorial east-central Pacific increased during April and May, when normally temperatures decrease in this region. During August - November ocean surface temperatures were at record monthly levels in the central and eastern equatorial Pacific. Departures from normal exceeded +4 C along the equator east of 140 W, and were greater than +5 C between 110 W and the Galapagos Islands. Over the past few seasons the NCEP statistical (CCA) and coupled model predictions have consistently indicated the development and persistence of strong warm episode conditions. The latest NCEP forecasts indicate that strong warm episode oceanic conditions, comparable to those observed during 1982-83, will continue into the boreal spring of 1998.

During warm (El Niño) episodes abnormal patterns of temperature and precipitation develop in many regions of the globe. These patterns result from changes in the distribution of tropical rainfall and the effect these changes have on the position and intensity of jet streams and the behavior of storms outside of the tropics in both the Northern and Southern Hemispheres. Current El Niño-related impacts include: 1) Wetter than normal conditions over the equatorial central and eastern Pacific (June-November); 2) Drier than normal conditions in Indonesia (June-November), most of Central America and Mexico (June-September), and equatorial South America east of the Andes mountains (June-November); 3) wetter than normal conditions over central Chile (May-August) and central portions of South America (October-November); and 4) exceptionally heavy rainfall over

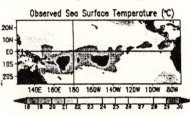
Observed Sea Surface Temperature (°C)

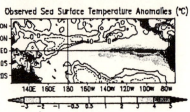
Observed Sea Surface Temperature Anomalies (°C)

7-day Average Centered on 31 December 1997

http://nic.fb4.noaa.gov/products/analysis_monitoring/ensostuff/index.html   1/8/98

FIGURE 5-2.   An example of the ENSO Advisory and climate forecasts issued by the Climate Prediction Center of NOAA. This advisory was issued on December 19, 1997.

Fifty-two of the seventy users (74 percent) believed in the accuracy of the seasonal forecasts when they became available. The other eighteen users reported that it took days or weeks before they came to accept the forecasts and their likely accuracy. Most of the twenty-nine individuals who "examined or considered" the forecasts, but eventually never applied them in formal internal decisions, were never able to develop confidence in the forecasts' accuracy or could not convince upper management to allow their use. One reason given for not initially believing the forecasts included the fact that different forecasts available presented different outcomes, a problem discussed in chapter 4. Forecasts that related the 1997–1998 event to the 1982–1983 event helped several decision makers resolve their uncertainties, particularly in regions where the 1982–1983 event had created damaging weather.

Two-thirds of the seventy interviewees (47 individuals) who said they used or considered the seasonal forecasts gave detailed answers regarding the applications of the forecasts. Forty-one used the forecasts to make explicit decisions, and the six others with detailed responses had examined the forecasts, assessed their impact in a decision model, and finally decided not to integrate the predictions into a decision.

Thirty-six of the forty-one users reported they had no difficulties in getting upper management to accept the use of the forecasts in one or more decisions. The forecast information was implemented in various institutional applications during the autumn of 1997 in thirty-five (79 percent) of the cases, and the forecasts were used in winter-season decisions in eleven cases and in summer 1997 decisions in five cases.

Most (90 percent) of the users believed that there was sufficient lead time prior to making the decision to incorporate the forecast information. Analysis of the kinds of decisions in which the forecasts were used revealed that thirty-three were operational decisions, and thirty-one were planning decisions. Some twenty-one decision makers used the climate forecasts in both types of decisions.

Sixty-six percent, or twenty-seven of the forty-one individuals who said they used the forecasts in one or more decisions reported that there had been a beneficial outcome. Only two (5 percent) said that using the forecasts had caused a negative outcome. Fourteen (34 percent) were not able at the time of the interview to indicate whether there had been a positive or negative outcome.

The forty-six persons who did not use the forecasts in formal action decisions were asked to provide reasons for not using the forecasts. Ten reasons were identified, and the number of persons reporting each is shown in Table 5-4. Most of the seventeen who immediately chose not to use the forecasts and the twenty-nine who examined them and eventually did not incorporate them into action decisions provided more than one reason; hence, the number of responses in Table 5-4 adds up to more than forty-six. The reasons provided were found to be similar to those identified in previous assessments of climate forecast uses by power utility decision makers (Changnon et al., 1995) and agribusiness users (Changnon, 1992).

The primary reason for not incorporating the climate forecasts into decisions was a lack of confidence in the forecast's accuracy. Figure 5-3 humorously illustrates the uncertainty many people felt about using the forecasts. Several individuals also found the information difficult to interpret or incorporate in their decision models. Importantly, several were "risk averse" and believed that they

TABLE 5-4.   Reasons given by interviewees for not using the El Niño–based seasonal climate forecasts.

Reasons	No. of responses
Lack of timely information	1
Did not believe forecasts	18
Level of accuracy/probabilities not given	22
Information difficult to interpret	11
Information not available when needed	1
Economic risks of failure too large to justify use	16
Differences between forecasts' contents	29
Forecasted conditions did not impact their regions	12
Did not know how to incorporate in their decision	2
Impacted by weather events not seasonal factors	9

could not risk using uncertain information in important decisions that would have very costly and potentially very negative outcomes to their institutions. Also, several decision makers were most interested in the occurrence of weather events, such as the number of storms, and the seasonal forecasts did not include such information (see chapter 2).

Seventy-one of the eighty-seven interviewed responded to the question about whether their job activities or those of their organizations, might change as a result of the highly accurate seasonal forecasts based on the El Niño 97–98 event. Some 41 percent (twenty-nine individuals) answered yes; 33 percent (twenty-three) answered no; and 26 percent (nineteen) were not sure what the future held. Some reported that they would now begin incorporating information from long-range forecasts because they believed that other organizations in their sector, that is, their competitors, would be using them.

At the end of the questionnaire, all eighty-seven persons were asked if they had any final comments relating to El Niño 97–98, climate forecasts in general, and

FIGURE 5-3.   This cartoon illustrates the uncertainty among the public, as well as among many weather-sensitive decision makers, about making decisions using the El Niño forecasts. (Reprinted with permission of Ed Stein, courtesy of the *Rocky Mountain News*)

the activities of CPC and other forecasting sources. Their comments and questions varied widely among sectors, but six interesting themes emerged from the discussions. These themes are listed in Table 5-5 along with the number of times a respondent raised the theme. Many of the themes relate to respondents' expectations for future forecasts. Clearly, increased interest in climate and its forecasting is indicated, as well as issues forecasters need to consider as additions to the forecasts.

HOW DECISION MAKERS IN DIFFERENT SECTORS REACTED

The data for decision makers in the major sectors was assessed to gain insight about possible sectoral differences and their reasons. Table 5-6 shows the breakdown, by sector, of the operational and longer-term planning applications of the seasonal forecasts. Some organizations used the climate forecasts for both types of activities.

Water Resources

Fourteen of the twenty-five water resource decision makers interviewed indicated they used the long-range climate forecasts. Eight considered using the forecasts but eventually decided against doing so for action-oriented decisions. Three said they did not use the forecasts at all. All individuals sampled performed water resource activities for various federal and state agencies. The primary source of the El Niño–based forecasts for these individuals was the official NWS forecasts issued by CPC or provided to them by some other government agency. Thirteen of the twenty-two (59 percent) who used or examined the climate forecasts said they believed the forecasts when they were presented to them.

Most (twelve of sixteen) water resource managers did not find it difficult to get organization management to accept their proposed use of the forecasts. Furthermore, a majority of them used the forecasts during the fall of 1997 (three others in the summer 1997 and two in winter 1997–1998). All fourteen users

TABLE 5–5. Themes summarizing the final comments presented by the 87 interviewees.

Comments/questions/concerns	No. of responses
The forecasts improved climate forecast credibility	31
Concern about quality of future forecasts and La Niña	8
Concerns about whether future forecasts would have the same level of accuracy	20
Lack of information (from CPC) at key decision times	9
El Niño was an interesting weather subject	22
Need more detailed storm-type information incorporated with the climate forecasts	18

TABLE 5-6.   Number and types of decisions based on
the El Niño forecasts for each major sector.

Group	Operational	Planning
Agribusiness	3	0
Power utilities	11	10
Water resources	9	12
Climatologists	3	2
Others (e.g., insurance)	2	4

believed that they were given sufficient lead time to incorporate the information into decisions.

The forecast information was used to make nine operational decisions and twelve long-term planning decisions. The types of operational activities influenced by the El Niño–based forecasts all included preparations for large amounts of rainfall expected:

- In California, where above-average winter precipitation was predicted, several water managers we interviewed had their staffs and members of the local community clear debris and other obstructions from ditches, canals, and river channels to allow greater volumes of water to flow, thus reducing the potential for localized flooding.
- In southern Utah, where above-average winter precipitation totals were expected, a decision was made not to enhance precipitation through use of an operational weather modification project.
- In Florida water levels in lakes and canals were reduced so that these systems could more effectively store the expected above-average precipitation and reduce the flood risk.

The forecasts were used for long-term planning for several water management schemes, including these:

- Western water managers developed summer 1998 water distribution schedules (such as appropriate reservoir levels) for the expected above-average spring runoff in the southern Rockies and below-average streamflow in the northern Rockies. These plans not only had to consider the threats from flooding but also had to take into account needs for hydroelectric power generation and environmental concerns, including the salmon migration in the Pacific Northwest.
- In Florida, water resource managers used the forecasts in developing a multimonth strategy to help a threatened bird species that breeds and develops in wetlands.
- In Florida and California, planning included the movement of large quantities of sand bags to regions where the predicted above-average precipitation was expected to swell streamflow to flood levels.

- All agencies made efforts to educate staff members and, often, to inform the local and regional water-concerned public about the potential problems likely to be created by the El Niño–based weather anomalies.
- In the Midwest, where the El Niño–related forecasts were for below-normal precipitation, water managers used the forecasts to establish plans for long-term river control operations based on the lower river and lake levels expected during 1998.

Nine of the fourteen water managers who used the forecasts reported that the use had provided major benefits. None said that the use had caused a negative outcome. Five managers did not yet know at the time of the interview whether the eventual planning outcome would be positive or negative. None of the nine who considered use of the forecasts to be beneficial reported any economic gains because most of the activities were preventive measures that succeeded in reducing flood-related losses. However, these decision makers considered that the outcomes were very positive and likely had large economic and societal values. In Florida, the bird species of interest did survive as a result of the decisions made by the water resource managers.

The three managers who did not use the forecasts (even after examining them) reported that there were three main reasons for nonuse: (1) they did not believe the forecasts (had little trust in them), (2) the forecasts did not rise to the level of accuracy desired, and (3) the use posed too large an economic and or environmental risk. Seven of the twenty-five water managers felt that their future job activities could change as a result of the accurate forecasts made of the El Niño 97–98 event. Nine believed that there would be no future changes (mainly due to restrictive rules that guide decisions in government agencies), and four did not know what might happen.

Agribusiness

Only three of the seventeen agribusiness decision makers interviewed reported they had used the forecasts. Seven said they did not use them, and seven reported that they considered the forecasts but in the end did not incorporate them in any action-oriented decisions. This low percentage of use (18 percent) was largely related to the lack of farming-related decisions made during the cold season (the period covered by the El Niño forecasts).

Those in agribusiness relied less on CPC for forecast information than did those in water resources or utilities. Most got the forecasts from university scientists and from others in their organizations. Ten reported that they had believed the climate forecasts when they were initially presented.

All of those who used the forecasts indicated that it was not difficult to get their management to accept their use. In all three cases, the forecasts were used in the fall 1997, and there was sufficient lead time to incorporate the forecast into decisions. The information was used in three operational decisions (Table 5-6). The types of decisions in which the forecasts were used included:

- Altering the type of crops to plant in areas of California where above-average precipitation was expected
- Developing irrigation schedules in California and Colorado that were based on the winter and spring precipitation forecasts
- Determining how many cattle could be fed and raised on the basis of grain availability in Montana (taking into consideration the forecast drier-than-average conditions).

All three cases of usage provided a beneficial outcome by reducing losses. However, none of the users was able to specify the monetary gains resulting from their actions.

The primary reason for nonuse of the climate forecasts by agribusiness managers was that their concern focused largely on summer conditions. One agribusiness manager stated, "There is very little we could do with an accurate cold season forecast—we are much more interested in the growing season." Those who could have used the cold season forecasts but chose not to reported three reasons: (1) they did not believe the forecast, (2) the information was considered difficult to interpret for their specific regions, and (3) the economic risk of use and possible failure was considered too great. Asked the question "Could future activities change due to the accurate seasonal predictions associated with the El Niño 97–98 event?" four said yes, six said no, and seven said they were uncertain.

Power Utilities

A relatively large number of power utility managers who deal with weather-sensitive conditions used the long-range El Nino–based forecasts compared to managers in other sectors. Fourteen of the twenty-two (64 percent) power utility officials interviewed indicated that they used the forecasts in making decisions. Three respondents indicated no use, and five said they considered their use for a time before ultimately deciding not to employ them. This relatively high rate of usage is linked to the strong need for information about future winter conditions.

Forecasts used by utility managers generally came from CPC and/or private meteorological forecasting firms employed by their companies mainly to provide short-term, one-hour to ten-day forecasts. Fourteen of those interviewed believed the forecast information when it was initially presented to them. Figure 5-4 shows the official probability forecast patterns issued by CPC for temperatures and precipitation in 2 different three-month periods.

Nine of the fourteen forecast users indicated they had no difficulty in getting top management to accept their desire to use the forecasts in various decisions.

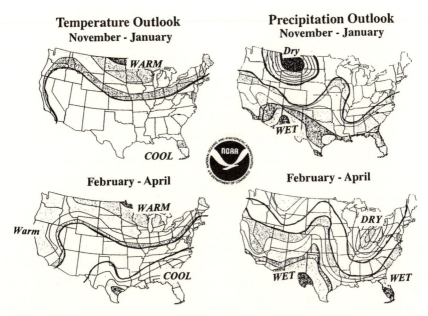

Temperature Outlook
November - January

WARM

COOL

Precipitation Outlook
November - January

Dry

WET

February - April

WARM

Warm

COOL

February - April

DRY

WET

WET

FIGURE 5-4. Forecast patterns involving probabilities of occurrence (the darker areas have the highest probabilities) issued by the National Centers for Environmental Prediction. Shown are the 3-month temperature outlooks (left) and the precipitation outlooks (right) for November 1997–January 1998 (top maps) and for February–April 1998 (lower maps). The forecasts indicate warm conditions across the northern half of the United States in both periods and wet conditions in the Southwest, the Deep South, and the Southeast. These patterns compare favorably with the actual outcomes (see Figure 5-5). The lightest shading (next to the heavy lines) indicates a probability of anomalies from 0 to 0.05; the next darkest shading represents a probability of from 0.05 to 0.1; the next from 0.1 to 0.15, and so on. Climate probabilities of equal chances for any outcome exist in the white areas.

Although the forecast information was implemented most frequently in the fall of 1997, forecasts were employed twice in decisions during the summer of 1997, and in four cases forecasts were employed during the winter of 1998. All but one utility manager felt that the climate forecasts provided sufficient time for them to incorporate the information into decisions.

The climate forecast information was used in eleven operational and ten long-term planning decisions (Table 5-6). The operational decisions made by managers in utilities distributed across the U.S. included:

- Altering maintenance and work schedules for coming months
- Making short-term decisions on natural gas purchases in the spot market
- Making decisions relating to the trading and purchasing of power

Planning decisions reported included:

- Determining how much coal to purchase for the 1998 summer season
- Predicting spring 1988 runoff amounts (hydro power applications)
- Identifying other utilities to trade with or sell power to in summer 1998
- Determining when to close plants for spring or summer maintenance.

Ten reported that the uses of the El Niño–related forecasts provided a beneficial outcome, while 4 did not yet know at the time of the interview whether the outcome was positive or negative. Estimates of the economic benefits to the utility resulting from forecast usage included $200,000 at a Midwestern utility, approximately $500,000 at two utilities (one located along the East Coast and the other in the northern Rockies), and more than $3 million for a southwest U.S. utility. A utility in the High Plains that used the forecast in deciding to wait and buy gas during the winter (and at much lower prices than in the fall) reported that the resulting savings to their customers were $30 million. Some of the sampled decision makers in the power utilities were unwilling to report the economic benefit of the corporate decisions.

Those utility officials who did not use the climate forecasts gave the following reasons for their nonuse: (1) they did not believe the forecast, (2) they questioned the level of accuracy, (3) the information was hard for them to interpret, and (4) El Niño–based forecasts did not impact their region of concern.

Fifteen utility officials (70 percent of the sample) responded positively to the question "Could future activities (job or organizational) change as a result of accurate long-term predictions associated with the 1997–1998 El Niño?" This was the largest percentage found among the three major sectors sampled (water resources, power utilities, and agribusiness). This indicates that organizations that are highly sensitive to winter weather conditions are more willing to incorporate uncertain forecast information in their decisions.

Climatologists Who Advise Selected Users

Included in the survey were eleven climatologists who advise specific individuals in weather-sensitive institutions. Three of the eleven climatologists reported they had recommended that a specific user they served should integrate the El Niño–based forecasts into its decisions. Six climatologists recommended the customer should assess the forecasts and reach its own decision about use. Many reported they did not emphasize using the long-lead forecasts because they were unsure how the forecasts might ultimately be used by the decision maker they served. Some expressed concerns over the amount of uncertainty associated with the forecasts and therefore did not recommend usage without great care. A few climatologists did not advocate using the forecasts because the predicted conditions did not affect the interests they served in their regions of responsibility. All the climatologists relied on the official CPC forecasts. Figure 5-5 illustrates the outcomes for temperatures and precipitation for the January–March 1998 period, and comparison with the forecast patterns (Figure 5-4) reveals many areas of agreement and some disagreement.

Five persons served by the climatologists, all working in transportation or emergency management agencies, reported that they had incorporated the forecasts supplied by the climatologists. They reported no problems convincing upper management about their wish to use the information. All five users reported they had adequate lead time in the fall 1997 to make timely decisions. The forecasts were used in three operational decisions (all relating to preparations for potential flooding) and two planning decisions (relating to plans for highway snow and ice removal needs). All users reported benefits from using the forecasts, but none was able to assign a monetary value.

The reasons offered by the six other persons advised by the climatologists who did not use the long-lead forecasts included (1) the low level of accuracy, (2) the difficulty of integrating the information into their decisions, (3) the high risk of economic failure, or (4) the failure of the forecasts to address their regions of the country. In response to the question about whether the successful 1997–1998 forecasts would alter future usage, three customers said yes, two said no, and seven said they were uncertain.

Only two of the eleven climatologists believed the El Niño forecasts improved the credibility of CPC and of the National Oceanic and Atmospheric Administration (NOAA). Several mentioned that concerns were raised by their users, including (1) the accuracy of future forecasts, (2) the lack of information at key times, and (3) the need to provide more detailed weather information in the forecasts. Overall, the climatologists believed that the amount of media attention devoted to El Niño 97–98 helped raise the public's interest in weather and climate and thus provided a unique opportunity to further educate the public about many weather-related topics (Figure 5-6).

Other Sectors

Interviews were conducted with individuals in the other major sectors previously described. Twelve individuals were interviewed, three from the weather insurance industry, five representing state and federal transportation agencies, and four from state emergency preparedness agencies. The sample size in these sectors was considered too small for a separate analysis of each sector, and they were grouped in the "other" category.

Five of the 12 used the forecasts to make decisions, three carefully considered the forecasts before rejecting their use, and four decided against use from the beginning. Half of the 12 reported they believed the forecasts when they were initially presented, whereas the other two individuals took several weeks

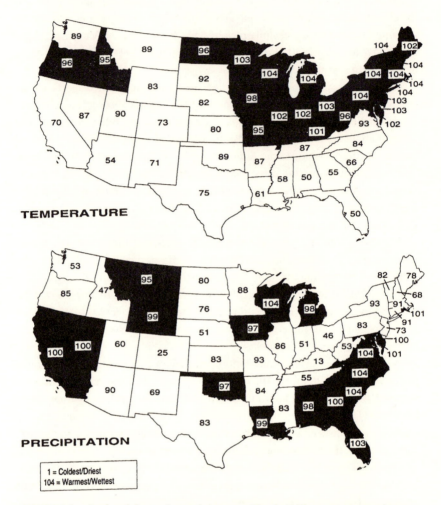

1 = Coldest/Driest
104 = Warmest/Wettest

FIGURE 5-5. Ranks of the each state's January–March 1998 temperature values and precipitation totals. Large numbers represent high temperatures or precipitation. The value of 104, the rank of the Michigan temperature, indicates that this was the warmest period in the state's 104-year record. These patterns reveal that three-month temperatures in the northern United States were exceptionally high (ranks in the 90s or higher) and that the precipitation totals in the Southeastern states and in California and Nevada were near record high values. These outcomes agree with the predicted values for these areas (Figure 5-4), but the wet outcomes in the northern Midwest (it was the wettest January–March period ever in Wisconsin) do not agree with dry predictions for dry weather in that area, as shown in Figure 5-4. (National Climatic Data Center)

During the start of the El Niño 97–98 event, a Northern Illinois University (NIU) climatologist worked with the university's heating plant manager to determine whether climate information and forecasts could assist him in making NIU's decisions on how and when to buy natural gas purchase each fall (Changnon et al., 1999). Purchases are normally made in advance of the cold season at market prices.

Information based on local climate conditions, a comparison of past natural gas usage and winter temperatures, and a model based on the equatorial Pacific sea-surface temperature (SST) El Niño/Southern Oscillation (ENSO) was provided to the manager well before the purchase decision had to be made. The model outlooks projected that the local winter had a 60 percent chance of having much-above-normal temperatures. The manager used this information and the CPC's long-lead forecast for winter, which also called for above-normal temperatures, to make the 1997–1998 gas purchase decision. He decided to "ride" the natural gas market, purchasing during the winter, as needed, rather than go with an early fall fixed-price purchase for winter. After the winter ended, he calculated that the gas cost $500,000 less than he would have paid had he purchased the gas in the fall. These savings became a key factor in the university's decision to employ a full-time applied meteorologist to assist various university administrators with weather-sensitive operational and planning decisions.

FIGURE 5-6.   A case study of forecast use.

or months to assess the forecasts and then came to believe them when the fall weather proved their accuracy. The sources of the forecasts included CPC for ten individuals; the private sector ranked second. Three individuals reported that the state-scale outlook maps issued by CPC (see Figure 5-7) had been very helpful in allowing them to assess potential regional outcomes for their areas of responsibility. Top management blocked requests to use the forecasts in four cases.

The forecasts were used in two operational situations: emergency preparedness activities in California and purchase of winter transportation supplies in Illinois. The forecasts were also employed in 4 planning decisions. These included:

- Developing plans to move people at risk in case of flooding (in two states)
- Planning to purchase varying amounts of road salt in response to winter precipitation forecasts
- Planning to alter insurance rates in regions expected to be adversely affected by stormy weather.

The four users from transportation and emergency preparedness sectors indicated that the forecast-related decisions had been very beneficial. Although they were unable to specify a monetary gain, they felt the outcomes had clearly reduced the potential for loss of life and property damage. Two of the three insurance persons interviewed indicated that incorporation of the available fore-

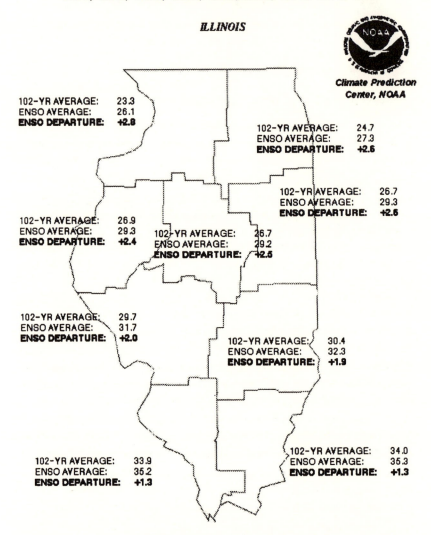

102-YEAR [1895 – 1996] NORMAL vs. ENSO-AVERAGE TEMPERATURE (°F)
BY CLIMATE DIVISION
DECEMBER - FEBRUARY
1940-41, 1957-58, 1965-66, 1972-73, 1982-83, 1986-87, 1987-88, 1991-92, 1994-95

ILLINOIS

Climate Prediction
Center, NOAA

102-YR AVERAGE:	23.3
ENSO AVERAGE:	26.1
ENSO DEPARTURE:	**+2.8**

102-YR AVERAGE:	24.7
ENSO AVERAGE:	27.3
ENSO DEPARTURE:	**+2.6**

102-YR AVERAGE:	26.7
ENSO AVERAGE:	29.3
ENSO DEPARTURE:	**+2.6**

102-YR AVERAGE:	26.9
ENSO AVERAGE:	29.3
ENSO DEPARTURE:	**+2.4**

102-YR AVERAGE:	26.7
ENSO AVERAGE:	29.2
ENSO DEPARTURE:	**+2.5**

102-YR AVERAGE:	29.7
ENSO AVERAGE:	31.7
ENSO DEPARTURE:	**+2.0**

102-YR AVERAGE:	30.4
ENSO AVERAGE:	32.3
ENSO DEPARTURE:	**+1.9**

102-YR AVERAGE:	33.9
ENSO AVERAGE:	35.2
ENSO DEPARTURE:	**+1.3**

102-YR AVERAGE:	34.0
ENSO AVERAGE:	35.3
ENSO DEPARTURE:	**+1.3**

FIGURE 5-7.   The winter temperatures for 9 regions in Illinois, showing the 102-year averages and the values when ENSO (El Niño) events occurred. These show that ENSO events brought temperatures between 1.3°F (southern area) and 2.8°F (northwest region) above the long-term average regional values. These "climate outlooks" from the Climate Prediction Center were available on the Internet during El Niño and were considered highly useful by decision makers who dealt with regional situations.

casts, regardless of the level of accuracy, would provide no benefit to their types of weather-related decisions.

The seven nonusers in this group gave standard reasons for rejecting use. Primary among these was the fear about forecast accuracy, particularly among upper management, and the lack of ways to integrate the information into their normal decision-making processes. Four of the twelve believed that the successful El Niño–related forecasts would have an effect on future interest in the use of long-lead forecasts.

SUMMARY AND RECOMMENDATIONS

Assessment of the uses of the El Niño–based seasonal forecasts for 1997–1998 by weather-sensitive decision makers in agribusiness, power utilities, water resources, and other sectors provided information about the level of usage, the gains and losses resulting from uses, those most likely to use future forecasts, and ways to improve or enhance the development and dissemination of the long-range climate forecasts. Responses obtained ranged from those expected at the start of this assessment (e.g., many could not use the forecasts because of management limitations) to the unexpected. For example, many of those who applied the forecasts to operational situations were involved in large economic decisions relating to the cold season.

Forty-one (47 percent) of the eighty-seven individuals interviewed indicated that they used the forecasts in one or more decisions; another twenty-nine said they examined, evaluated, and/or considered implementing the forecasts in decisions before finally choosing not to use them; seventeen said that, although they were aware of the El Niño–based forecasts, they did not consider their use. The reported level of usage (47 percent) was much higher than the 32 percent level of use found in a similar assessment, in 1996–1997, of usage of the new type of long-range forecasts that had been issued by CPC since January 1995 (Changnon, 1997). Importantly, 66 percent of the sample (twenty-seven of forty-one users) reported the decisions based on the forecasts had led to beneficial outcomes, and only 2 percent (two individuals) reported a negative outcome. The others were unable to give an evaluation at the time of the interviews, mainly because evaluations of planning-based outcomes had not been completed.

The power utility officials we sampled were most likely to use the forecast information. Agribusiness decision makers were found least likely to have incorporated the forecasts in a decision, primarily because most of their important weather-related decisions pertain to the growing season and the El Niño–based forecasts focused primarily on fall 1997, winter 1997–1998, and early spring 1998.

Most (forty-one of forty-seven) of those who used the forecasts indicated that their management had accepted their recommendations to use the forecasts. A large majority indicated that they used the forecast information in their fall 1997 decisions (prior to winter) and that having a season (or approximately three months) of lead time was adequate for interacting with management and implementing decisions. Decisions made in the power utility sector were nearly evenly split between operational and long-term planning applications, whereas water resource managers placed more emphasis on using climate forecasts in long-term planning decisions. Many water resource managers indicated that most of their operational decisions were based on weather events, such as number of heavy rain periods, which were not predicted in the climate forecasts.

Findings and Recommendations

One of this study's goals was to assess the use of CPC's climate predictions and to determine how various other sources of forecasts affected usage of the CPC data. Sixty-one, or 87 percent of the seventy individuals who used or considered using the El Niño–related forecasts assessed both the official forecasts from CPC and forecasts from one or more other sources. These other forecasts often provided information that interviewees perceived as inconsistent with expected regional seasonal conditions. As a result, many who considered the use of the forecast information were confused and decided against using it.

Finding 1: Inconsistent/differing forecast information often leads to nonuse of forecasts. We therefore recommend that the federal government continue conducting seasonal (fall and spring) meetings across the country to describe the upcoming official forecasts and their potential impacts. NOAA should identify individuals (on the Internet and elsewhere) who provide only the official forecasts information and who can aid decision makers who may not understand how to deal with conflicting climate forecast information or the reasons for the differences. Efforts should include seeking more precise terminology and ways to present forecast information in a numerical format.

CPC and other scientific organizations released information about the impending El Niño and its influence on future weather conditions to the media early (in summer 1997). Most of those we interviewed (74 percent) said the forecast information was available in time for their decisions, and these decision makers acted soon after receiving the forecasts. Many who did not initially believe the El Niño–based seasonal temperature and precipitation forecasts waited until the fall or winter weather conditions developed and proved the accuracy of the predictions before deciding to employ the forecasts in a decision.

Finding 2: Some potential users are proactive, whereas others need convincing and are more reactive to the accuracy of current forecasts. We suggest that NOAA be alert to the need to distribute information widely when ENSO conditions develop in the future. Many who considered themselves "reactive" during the El Niño 97–98 event indicated that they would be better prepared to take a proactive stance when the next El Niño occurs. Discussion of the accuracy of both past and current forecasts is highly desirable and is of great value. The applied climatology community should work closely with decision makers who are uncertain and reactive to identify and develop types of enhanced information, including decision models that can be used with climate forecasts to improve decisions (Changnon, 1998).

Two-thirds of the 41 individuals who implemented the forecasts in a decision indicated that the predictions led to some benefit for their organization. Economic benefits reported were all from utilities, and gains ranged from $200,000 to $30 million. A third of those who used the forecast information said they did not know at the time of the interviews whether there had been a positive or negative outcome.

Finding 3: Those most willing to use the El Niño–based seasonal forecasts and those able to identify financial outcomes were from power utilities. We therefore suggest that NOAA and the private meteorological sector may want to expand their interactions with users in this sector. Most water resource managers are employed by the government, and many tend to be risk averse when it comes to using uncertain forecasts in operational decisions. The sampling also revealed that most agribusiness applications of long-lead forecasts pertain to growing season conditions, and the highly accurate cold-season forecasts based on El Niño conditions have few applications.

The reasons given for nonuse of the information, even after the forecasts were examined and carefully considered, fell into ten categories. Results agreed with those from prior research (Changnon, 1997; Changnon et al., 1995). The leading reasons for decision makers' nonuse of El Niño forecasts were that:

- They did not believe the forecasts
- They considered the accuracy level of the forecasts to be unknown or not high enough for their needs
- They felt the forecast-related information was too difficult or too subjective to interpret
- They believed the economic risks of a bad decision were too large
- They concluded that the forecast climate anomalies did not impact their region.

Agribusiness and insurance officials sampled were most likely not to use the forecast information.

Other reasons for nonuse included the following:

1. Inconsistent forecast information was obtained from various sources.
2. Users did not know how to incorporate probabilistic information in their decisions.

3. Organizations were prepared for any climate anomaly that might occur.
4. Weather events posed larger problems for the organization than the forecast seasonal conditions.
5. Seasonal forecasts did not impact the types of decisions interviewees were involved in.
6. Many decision makers were restricted by organizational rules and government regulations.

Finding 4: Although the El Niño–based forecasts were correct for many U.S. areas, numerous problems prevent the use of long-term forecasts in many weather-sensitive decisions. NOAA and other applied climatologists who are interested in seeing the climate forecasts go from development-and-dissemination to implementation in cases where uncertainty can be handled need to actively work with users to inform them and to help them understand the forecast information and uncertainty. Bridging the "information gap" that currently exists between climatologists, who understand climate data, products, and forecasts, and weather-sensitive decision makers who do not understand climate information or how to incorporate it into decisions is a fundamental task that needs to be addressed before the wise use of climate forecasts can significantly increase (Changnon, 1998).

Decision makers were asked whether the highly accurate seasonal forecasts based on the El Niño 97–98 event would alter how they did their jobs in the future. Interestingly, 41 percent said that it would have some influence on how they made future decisions, especially when El Niño or La Niña conditions were influencing the atmosphere and were integral to the forecast conditions. Nearly all twenty-two power utility officials sampled agreed that this 1997–1998 event had altered how they viewed and would use climate forecasts. Their decisions to incorporate ENSO-derived climate forecasts are tied to economic pressures facing the power industry as a result of deregulation. Individuals in this group were the only ones who could identify economic outcomes within a few months after the application. Conversely, water resource managers reported that the success of the 1997–1998 forecasts would have little or no impact on what they did in the future. Their reasons were that: (1) they had always been aware of long-lead forecasts and basically did not trust their accuracy (and one correct forecast would not change that perception), (2) rules and regulations within their public organizations (e.g., Army Corps of Engineers, Bureau of Reclamation, and other government agencies) limit the use of climate forecasts, and (3) the economic and environmental risks associated with the use of uncertain information in key decisions were too high.

Finding 5: Money talks! The private meteorological sector should focus its efforts on the promising sectors and develop specialized, more detailed climate information to enhance the climate forecasts. Those who used the 1997–1998 forecasts employed them to aid in making timely economic decisions.

The eighty-seven individuals sampled in this study were asked whether they had any final thoughts about CPC's climate forecast, ENSO conditions, NOAA, and other related issues. Twenty-six said that the forecasts based on El Niño 97–98 had improved the credibility of CPC and NOAA. A similar number commented that the El Niño phenomenon was an interesting weather topic and that they had learned a great deal about it and about climate in general. Conversely, many expressed concern about whether the confidence level associated with the climate forecasts for the upcoming winter (1998–1999) would be as high as that for the 1997–1998 season. If it were not as high, interviewees wondered how CPC would express that uncertainty to decision makers in various organizations. Several mentioned that if the 1998–1999 seasonal forecasts were incorrect, most of the forecast credibility that had been gained from the 1997–1998 event would be lost. Many also wondered if more regional and storm-specific information could be developed for inclusion in the forecasts to meet the organizational needs of those working in water management (e.g.,variations of precipitation in a watershed) and for power utilities (e.g., temperature and heating degree day anomalies) for specific periods.

A premise when the study began was that most U.S. users would be located in regions that had been impacted by previous El Niño events (Ropelewski and Halpert, 1986), such as the West and the Southeast. Although that hypothesis proved generally true for water resource and agribusiness decision makers, we found that power utility officials were generally interested in the forecasts no matter where they were located. This finding may be related to the fact that natural gas and heating oil have been deregulated in the United States, and utilities have greater freedom to trade supplies throughout the country. Thus, utility decision makers located in a region not expected to be detrimentally impacted by adverse weather conditions have an opportunity to use the climate forecasts to identify impacted areas and to decide where their organizations should place an emphasis on trading. On the other hand, water resource and agribusiness managers are most concerned with weather in a specific region (e.g., a watershed, a crop-growing area, or a sales region), and their decisions are generally influenced primarily by weather events there.

Perhaps the most important outcome of this assessment is the discovery that most institutions (66 percent) in which individuals used the El Niño–derived climate forecasts experienced some form of benefit. Only 5 percent had a negative outcome, and 29 percent did not know the outcome at the time of the interviews. Those who did not use the highly accurate forecasts missed major opportunities. On the basis of the national media hype, which focused on negative impacts expected from the El Niño weather, few in the general public expected any positive benefits from this weather phenomenon, but the use of forecasts by utilities in timing their natural gas purchases produced huge benefits for consumers (see chapter 6).

Finding 6: Use of accurate seasonal forecasts yields largely positive outcomes. In today's information age, NOAA needs to exploit these findings when future forecast opportunities (ENSO conditions) exist to alert impacted sectors that can either minimize losses or maximize profit by using the forecast information. They also need to explain when the skill levels are lower and present the probabilities associated with the outcomes predicted.

REFERENCES

Brown, B. G., and Murphy, A. H. 1987. *The Potential Value of Climate Forecasts to the Natural Gas Industry in the United States.* SCIL Report 87-2. Corvallis, OR: Oregon State University.

Brown, B. G., Katz, R. W., and Murphy, A. H. 1986. On the economic value of seasonal precipitation forecasts: The fallowing/planting problem. *Bulletin Amer. Meteor. Soc., 67*, 833–841.

Changnon, D. 1998. Design and test of a "hands-on" applied climate course in an undergraduate meteorology program. *Bulletin Amer. Meteor. Soc., 79*, 79–84.

Changnon, D., Creech, T., Marsili, N., Murrell, W., and Saxinger, M. 1999. Interactions with a weather-sensitive decision maker: A case study incorporating ENSO information into a strategy for purchasing natural gas. *Bulletin Amer. Meteor. Soc., 80*, 1117–1126.

Changnon, S. A. 1982. Possible uses of long-range weather outlooks in water resources. *Proc. International Symposium on Hydrometeorology.* Denver, CO: American Water Resources Association, 231–234.

———. 1991. Applied climatology: Atmospheric sciences biggest success story faces major new challenges. *Preprints Seventh Conference on Applied Climatology.* Boston: American Meteorological Society, 1–3.

———. 1992. Contents of climate predictions desired by agricultural decision makers. *J. Appl. Meteor., 31*, 1488–1491.

———. 1997. *Assessment of Uses and Values of the New Climate Forecasts.* CRR-43. Mahomet, IL: Changnon Climatologist.

Changnon, S. A., and Changnon, J. M. 1990. Use of climatological data in weather insurance. *J. Climate, 3*, 568–575.

Changnon, S. A., Changnon, J., and Changnon, D. 1995. Uses and applications of climate forecasts for power utilities. *Bull. Amer. Meteor. Soc., 76*, 711–720.

Changnon, S. A., Sonka, S. T., and Hofing, S. 1988. Assessing climate information use in agribusiness. Part I: Actual and potential use and impediments to usage. *J. Climate, 1*, 757–765.

Glantz, M. 1987. Politics, forecasts, and forecasting: Forecasts are the answer but what is the question? In *Policy Aspects of Climate Forecasting.* Washington, DC: Resources for the Future, 81–96.

Pfaff, A., Broad, K., and Glantz, M. H. 1999. Who benefits from climate forecasts? *Nature, 397*, 645–646.

Ropelewski, C. F., and Halpert, M. S. 1986. North American precipitation and temperature patterns associated with El Niño/Southern Oscillation (ENSO). *Mon. Wea. Rev.*, 114, 2352–2362.

Sonka, S. T., Changnon, S. A., and Hofing, S. 1992. How agribusiness uses climate predictions: Implications for climate research and provision of predictions. *Bulletin Amer. Meteor. Soc.*, 73, 1999–2008.

Vogelstein, F. April 13, 1998. Corporate America loves the weather. *U. S. News & World Report*, 48–49.

6

 Impacts of El Niño's Weather

STANLEY A. CHANGNON

T he societal, economic, and environmental consequences of weather events and climate conditions in the United States vary across the nation as a result of hot and dry conditions in one region and cold and wet conditions in others, or storms in one area and none in others. Thus, for any given period, such as a season or year, the weather-caused impacts in the United States reveal a mix of winners and losers. This was certainly true with the impacts resulting from El Niño 97–98.

The official National Oceanic and Atmospheric Administration (NOAA) predictions issued in June 1997 calling for more storms in parts of the nation and heavy precipitation for the South and Far West (Climate Prediction Center, August 13, 1997) created major fears about large economic and social losses. The warnings of FEMA and the ensuing media hype created a nationwide perception that all "El Niño weather" was going to be damaging. This fear is illustrated in the cartoon in Figure 6-1. For example, the *Financial Times* (July 28, 1997) tied the strong El Niño 97–98 conditions to the huge U.S. losses due to El Niño 1982–1983, with 161 killed and losses of $2.2 billion. Such connections and citations resulted from the fact that the official El Nino predictions and Federal Emergency Management Agency (FEMA) warnings were comparing the large El Niño 97–98 to the large 1982–1983 event (CPC, July 1997; FEMA, August 12, 1997).

California newspapers also focused on the 1982–1983 losses in that state, which included fourteen killed and $265 million in damages (*San Francisco Chronicle*, August 14, 1997; *Sacramento Bee*, October 15, 1997). This helped create considerable concern and launched major mitigation endeavors in California, where storm and rain predictions were ominous. The resulting 1997–1998 mitigative activities in California reduced losses and were a major beneficial impact of the use of the long-range predictions of the Climate Prediction Center (CPC) and the warnings issued by FEMA that promoted mitigation actions.

The potential impacts resulting from issuance of the official predictions of a fall–winter–early-spring period of above-normal temperatures and below-normal precipitation for the northern sections of the United States were largely ignored by most scientists and the media, since these were not seen as creating negative impacts. However, a few scientists did identify some possible benefits, such as fewer Atlantic hurricanes and lower energy prices in the Northeast (Hall, 1997).

The role of the scientific community in focusing on negative, rather than positive, impacts from El Niño weather was an important part of the "bad outcome" theme that surrounded El Niño and permeated the news media during 1997 (Glantz, 1998). Institutions also played a key role. For example, FEMA issued a series of news releases, best labeled as warnings, during August-October, and these focused on dangers and damages in an effort to draw the public's attention to the need for mitigative efforts in the West and South (FEMA, August 12 and September 12, 1997). A scientific report prepared in October for the insurance industry predicted several bad El Niño outcomes, including excessive flooding in the West, Southwest, South, Southeast, and Central Plains (Skinner et al., 1997). The director of the U.S. Geological Survey, in testimony before Congress in October 1997, predicted more flooding and increased water quality

FIGURE 6-1.   This cartoon, published in August 1997, captures the public's sense of deep concern and alarm over the bad weather conditions being predicted by atmospheric scientists during the summer of 1997 for the coming months for many parts of the nation. (Reprinted with permission of Ed Stein, courtesy of the *Rocky Mountain News*)

problems because of El Niño but failed to identify any positive outcomes derived from the additional water in the arid west (Shaefer, 1997).

A NOAA report that reviewed the El Niño–derived winter weather conditions and their negative impacts reflected the commonly held perspective that the impacts had been bad, as expected. The report stated, "The winter of 1997–1998 was marked by a record breaking El Niño event and unusual extremes in parts of the country. Overall, the winter was the second warmest and seventh wettest since 1895. Severe weather events included flooding in the southeast, an ice storm in the northeast, flooding in California, and tornadoes in Florida. The winter was dominated by an El Niño–influenced weather pattern, with wetter than normal conditions across much of the southern third of the country and warmer than normal conditions across much of the northern two-thirds of the country" (Ross et al., 1998). The report contained no mention of the numerous and sizable positive outcomes from the winter weather conditions in the north.

Without a costly major study, it is impossible to derive a reasonably precise measure of the economic and environmental impacts of a major nationwide event like the El Niño–generated weather conditions. A recent in-depth study identified the difficulties of precisely estimating losses caused by natural hazards (National Research Council, 1999). However, by using data in news accounts, business documents, and government reports, along with data on insurance losses from catastrophes, estimates of the impacts have been derived. On the basis of past studies involving extensive assessments of the economic impacts of natural hazards, it is likely that the estimates derived for the El Niño 97–98 impacts are between 15 and 30 percent of the true costs (Changnon, 1996; Guimares et al., 1993; West and Lenze, 1994).

This chapter first addresses the national impacts on human lives and the economy that resulted from weather events attributed to El Niño 97–98. The net economic effect was surprisingly positive, and less government relief was needed than in prior non–El Niño winters. The next section contains a detailed diagnosis of the impacts in the Midwest. This regional analysis illustrates in detail the types of negative and positive effects created by El Niño–generated weather.

NATIONWIDE LOSSES

El Niño–influenced atmospheric conditions created a considerable amount of damaging weather. In March, a leading NOAA scientist stated that El Niño 97–98 was "the most damaging ever" (Friday, 1998). As explained in chapter 4, the series of weather disasters from October 1997 through May 1998 was attributed to the record El Niño of 1997–1998 (Dole, 1998), and these weather disasters were noteworthy for their variety and for their wide distribution across the nation. Some debate exists about which weather conditions during El Niño were totally or partially due to (or enhanced by) El Niño's influence on the atmosphere (see chapter 4). In assessing losses, the events and weather conditions included as being El Niño–related were those that had been so identified by government atmospheric scientists acting in an official capacity.

As predicted when El Niño oceanic conditions grew to record proportions in the tropical Pacific during June–August 1997, California was assaulted by a series of coastal storms and heavy rains that caused floods, numerous landslides, and damage to the state's valuable agriculture, with losses totaling $1.1 billion statewide (Andrews, personal communication). The sector most adversely affected was agriculture, at $0.5 billion, or 3 percent of California's crop value for the year. Florida, Texas, and several other southern states were struck by several severe rainstorms and numerous tornadoes, events uncommon in winter, and these were attributed to El Niño (Le Comte, 1999). Tornadoes led to more than one hundred deaths, and El Niño–caused property and agricultural losses in Florida ultimately reached $500 million. In Florida, fifty-four of its sixty-seven counties were declared federal disaster areas (National Climatic Data Center, July 1998). A record early snowstorm swept across the High Plains and upper Midwest in October, and then extremely severe ice storms (Figure 6-2) struck the Northeast in January, creating losses in excess of $400 million and twenty-eight deaths (Ross et al., 1998). The intensity of both storms was attributed to El Niño (Ross et al., 1998; Wolter, 1997).

By the end of May 1998, the national death toll caused by weather conditions attributed to El Niño was 189. This total included forty-two deaths from Febru-

FIGURE 6-2.   Downed power lines from the massive ice storm in the Northeast early in January 1998, an event attributed to El Niño that caused damages of $400,000 in the United States and $5 billion in Canada. (Courtesy Hank Mueller)

ary tornadoes in Florida, twenty-eight deaths from the January ice storm and from storm-related floods in the Carolinas, seventeen deaths in California during the entire winter, two from a Minnesota tornado, three drownings caused by snowmobiling on thin ice in Michigan, twenty-four deaths from a February snow and rainstorm in fourteen eastern states, sixty-five deaths from tornadoes during March and April in southeastern states (Figure 6-3), and eight drownings in Texas from a December flood-producing rainstorm. President Clinton visited damaged areas of Florida and California late in February and stated, "The people of California and now Florida are giving the people of the U.S. some painful examples of the excesses of this El Niño, which is apparently the strongest ever in this century" (Clinton, March 1, 1998).

The property insurance industry identified fifteen catastrophes, events each causing greater than $25 million in insured losses, during the nine-month period ending in May 1998, when El Niño's influence on the weather had largely ended. The total insured property losses caused by these fifteen catastrophes was $1.7 billion (Property Claim Service, 1998). States where insured losses from three or more catastrophes occurred included Alabama, California, Florida, Georgia, Louisiana, Mississippi, and North Carolina (Figure 6-4). Forecasts calling for storms on the West Coast and in the Deep South were thus correct. Florida experienced losses in five of the fifteen catastrophes, and three of these caused more than $100

FIGURE 6-3.   A tornado outbreak on April 8, 1998, in Alabama and Georgia followed a three-day period of heavy rains. This photograph shows the severe tornado damage to a rural Alabama farm house and its surrounding trees. (Courtesy Charles Bright)

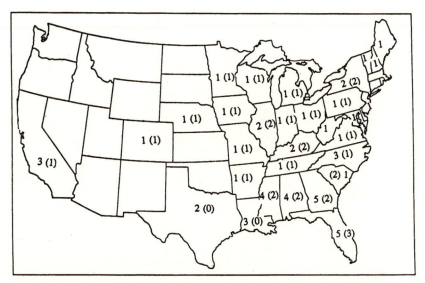

FIGURE 6-4. The number of catastrophes (defined as events causing national insured losses greater than $25 million) related to El Niño–generated weather conditions that occurred between September 1997 and May 1998. Shown are the number of times a catastrophic event occurred in each state (most weather catastrophes resulted in losses in several states). The values in parenthesis are the number of times that the national insured losses caused by catastrophic event were greater than $100 million. Florida led in both categories with five catastrophes >$25 million and three catastrophes >$100 million.

million in losses nationally. The single greatest insured storm loss was $305 million, the result of heavy rains and flooding, hail, and tornadoes in a storm system on April 15–17 that swept across Arkansas, Missouri, Kentucky, Tennessee, and Illinois (and killed eleven). The major winter storms in October and January account for the catastrophe counts (see Figure 6-4) in the states constituting the central High Plains, upper Midwest, and New England.

A severe drought developed in Hawaii as a result of El Niño's influences on the region's weather, depleting reservoir supplies, hurting certain crops, and leading to water use restrictions that lasted a year. The much-above-normal fall and winter precipitation in California and Florida flooded many fields and devastated several fruit and vegetable crops. National prices for fresh produce rose 7.9 percent in January, retreated in February, and then rose 5 percent in March. The floods and storms in California were cited as the main reason for a raise in the price of food of 0.4 percent in February (Department of U.S. Labor, March 14, 1998), and the media helped make the California flooding a national topic of discussion (Figure 6-5). Food processors also suffered from a lack of produce and complained of the poor quality of the fresh produce coming from California and Arizona (*Detroit News*, March 3). Prices for strawberries doubled, those for cau-

FIGURE 6-5. These headlines, from a variety of newspapers published during the January–June 1998 period, illustrate the variety of problems caused by El Niño weather conditions for tourism, agriculture, water supplies, home owners, and transportation.

liflower tripled, and the U.S. Department of Commerce announced that food prices rose 0.6 percent in May largely because of an 11.9 percent increase in vegetable prices (*Detroit News*, June 5). A USDA meteorologist reassured the public that the higher prices would retreat by June (Peterlin, April 15).

The tourist business, which is dependent on sunny and warm winter weather in Florida and the California coastal areas, was hurt by a 30 percent drop in tourists during the winter and early spring, although the skiing industry in California had much-above-average business, thanks to copious snow (*USA Today*, March 2, 1998). The national media extensively reported on the many problems in California (see Figure 3-5). Many ski resorts in the Midwest and the Northeast were hurt by the lack of snow; they faced increasing costs to make artificial snow, and poor conditions kept many skiers away. Michigan reported that income at the state's ski resorts had decreased by 50 percent (Pearce and Smith, 1998). Among the businesses most impacted negatively by El Niño–generated weather were providers of natural gas and heating oil (because temperatures were so mild in the northern United States); farmers who produced fresh fruits and vegetables in California and Florida and cotton in Arizona, places where it was too wet; manufacturers of snowmobiles and snow removal equipment, including shovels (because of the low snowfall in the northern United States); and producers of salt, who were victims of the low snowfall and the fewer-than-normal winter storms and hence saw sales drop (*USA Today*, March 2, 1998). Retailers in California and Florida reported 3 to 5 percent decreases in sales as a result of the cool and wet weather (*Wall Street Journal*, March 6, 1998). The wide variety of problems caused by the El Niño weather are illustrated by the headlines in Figure 6-5.

Impacts of El Niño's weather in other regions of the world also produced negative impacts in the United States. For example, the drought in Panama led to a lowering of the water level in the canal; this reduced shipping loads and increased costs for shipping for five months. Davis (1997) assessed many of these external impacts and showed how the drought in Central American hurt the production of vegetables exported to the United States, causing their prices to rise 10 percent for three months. Further, commodity traders dealing with Central American agricultural products did an extensive business. The El Niño–related drought in Southeast Asia cut production of coffee and palm oil, raising U.S. prices (Davis, 1997).

Even after the storm activity ended, more El Niño–related damages continued to occur. Widespread fires broke out in Florida during June, fueled by a heavy growth of underbrush caused by the unusually heavy El Niño–caused winter rains. In Florida and Texas, two states predicted to have above-normal rainfall in the spring because of El Niño, spring rainfall was well below normal and drought conditions developed, helping to create the Florida fires in June and greatly hurting the crops in both states (NOAA, June 1998).

FEMA relief payments for El Niño–caused storm losses reached $289 million by the end of March, but this was less than relief payments had cost in the two prior winters, which were not related to El Niño (Bunting, 1998). By the end of February 1998, thirty-five California counties had been declared federal disaster areas, and federal aid to California included funds for temporary housing, small business loans, the rebuilding of homes, levee repairs, payment of laborers for lost income, and repair of damaged roads (FEMA, February 22, 1998). Californians had responded to FEMA's early warnings about the potential for flooding, and the number of flood insurance policies grew from 265,000 in October 1997 to 365,000 by December 1997 (FEMA, March 4, 1998). Eighteen presidentially declared disasters were identified from fall 1997 through April 1998, and all were partly attributed to El Niño's influence on the atmosphere (Leetma, March

1998). El Niño events had become stronger and more frequent since 1980 (Hart, 1999), certainly one reason for the increased losses from weather-related natural disasters over the preceding fifteen years (Changnon et al., 1997).

In summary, the national economic losses that could be estimated included the following:

- Property losses totaled $2.8 billion. Insured losses were $1.7 billion, and noninsured losses were estimated as $1.1 billion, on the basis of the fact that insured storm losses normally represent 65 percent of all structural losses caused by storms (National Research Council, 1999)
- Federal government relief cost $410 million.
- State costs totaled $125 million.
- Agricultural losses were put at $650 to $700 million.
- Lost sales in housing and snow-related equipment reached $60 to $80 million.
- Losses in the tourist industry were estimated at $180 to $200 million.

The U.S. losses totaled about $4.2 billion and 189 deaths. *Science* reported in February 1999 that the global losses due to El Niño 97–98 included 23,000 lives and $33 billion in damages (Kerr, 1999).

NATIONWIDE BENEFITS

As noted, weather conditions across the United States for any given month, season, or year tend to produce losers and winners. The mild, almost snow-free winter in the northern United States produced by El Niño's influence on the atmospheric circulation over North America resulted in several major gains and some losses in the northern sections of the nation.

Many fewer lives were lost because of bad winter conditions (e.g., bad roads, low temperatures, severe winter storms) than normally occur. Estimates from various parts of the United States indicated a nationwide drop from an average of about 850 winter deaths to fewer than one hundred lives lost to winter conditions during 1997–1998. The mild, near-record-high winter temperatures of 1997–1998 (see Figure 1-3) meant few really cold temperatures, and this greatly reduced the lives lost to extreme low temperatures. Lives lost nationwide to extreme cold totaled thirteen (Parrish, 1999), compared to the annual average of 770 (Adams, 1997), a reduction amounting to 757 lives. Winter snowstorms and ice storms, fewer than normal, led to thirty-three deaths nationally (National Weather Service, 1997–1998), which is fourteen fewer than average (Kocin, 1997). Vehicular injuries and deaths resulting from winter season accidents also decreased (Pearce and Smith, 1998); the December 1997–March 1998 total nationwide was sixty-four deaths (National Highway Traffic Safety Administration, 1999), or fifty-seven fewer deaths than the average for the two prior winters (National

Safety Council, 1999). In sum, these national reductions mean 828 fewer deaths than in an average U.S. winter.

El Niño's influence on the atmosphere led to the elimination of major Atlantic hurricanes during 1997 (CPC, November 1997; Gray, 1997), and annual hurricane damages in the United States have been averaging $5 billion per year in the 1990s (Pielke and Landsea, 1998). This lack of hurricanes meant an enormous savings to home and business owners, the government, and insurers. It also meant no lives lost to hurricanes, which have caused on average twenty deaths per year since 1986.

The abnormal warmth led to major reductions in heating costs because of reduced use of natural gas and heating oil, as illustrated by the home heating bill shown in Figure 6-6. Nationwide, the energy savings were 10 percent (Ross et al., 1998), for a savings of $6.7 billion. Utilities using the climate predictions also bought natural gas and heating oil supplies at much lower prices during the winter, rather than having signed costlier early-season contracts, and this further reduced heating costs to their customers. Charges to consumers for natural gas fluctuate with the price—if the provider pays a high prices for gas supplies these are passed on to the consumer, whereas if gas is purchased at a low price, the consumer benefits. The major reduction in the use of natural gas and oils was sufficient to have an effect on global oil prices, and El Niño's influence in bringing warm seasons to North America and Europe was cited as one of three

	1997			1998 (El Niño)		
	ILLINOIS POWER			**ILLINOIS POWER**		
JANUARY	Residential Electric Service	$	245.49	Residential Electric Service	$	189.43
	Residential Gas Service	$	252.22	Residential Gas Service	$	141.02
FEBRUARY	Residential Electric Service	$	202.66	Residential Electric Service	$	171.37
	Residential Gas Service	$	172.44	Residential Gas Service	$	121.41
MARCH	Residential Electric Service	$	117.65	Residential Electric Service	$	110.10
	Residential Gas Service	$	86.87	Residential Gas Service	$	80.32

3-MONTH TOTALS				
	Electric	$566	Electric	$470 (83% of '97)
	Gas	$511	Gas	$343 (67% of '97)
	Total	$1,077	Total	$813 (75% of '97)

FIGURE 6-6. Comparison of residential bills for electric and gas service (at a Midwestern home heated partly by electricity but mainly by natural gas) for (1) the El Niño winter from January to March 1998 (with a mean temperature 7°F above normal) and (2) for the prior winter (1997), which had near normal temperatures (+1.2°F). The El Niño winter led to a 33% reduction in gas costs (31% in therms) and a 17% reduction in electricity costs (15% in kwh).

factors that led to a major reduction in gasoline prices that began in March 1998 (Stamper, 1998).

Not only were many fewer persons killed because of the mild, storm-free winter weather, but many people changed their normal winter behavioral patterns. Thousands went outdoors more, millions went shopping, many altered their types of recreation, and almost everyone enjoyed better health than in normal winters. There were notably fewer airline and highway transportation delays due to inclement weather, bringing less stress and increased profits, estimated at 5 to 8 percent, to major airlines and the trucking industry.

The lack of winter snowfall and freezing rain led to major reductions in the use of salt on highways and streets, saving money and minimizing environmental impacts. It also reduced expected overtime payments to street crews for snow removal and collectively brought major savings to state and local governments. For example, the savings reported in the Chicago metropolitan area totaled $21 million (Fonda, 1998).

The generally good weather, with little precipitation and temperatures averaging 9° F above normal, also had a major influence on construction, retail shopping, and home sales. Many retail chains reported record high sales for January–March. The U.S. Department of Commerce (March 1998) reported that the construction of new homes in February was up 6 percent from January, the highest monthly increase since November 1987, and that income and employment in the construction industry from December through February had increased 25 percent, raising workers' incomes an estimated $350 million over their incomes in recent winters (National Association of Realtors, 1998). Retail sales rose 4.9 percent above 1997 values in January (*Wall Street Journal*, February 1998), 5.7 percent in February (*Wall Street Journal*, March 1998), and 5.4 percent in March (*Wall Street Journal*, April 1998). A headline in the *Chicago Tribune* in early April announced "Retail Sales Soar in March" and was followed by the comment that record consumer spending had boosted the stock market. Record March sales, ranging from 4 percent to 12 percent above previous March levels, were achieved by a few retailers, including Kmart ($2.8 billion), Wal-Mart ($11.7 billion), and Sears ($3.7 billion).

A well-known economist noted that the strong economic growth of the first quarter of 1998 was caused largely by the unusually warm and dry winter, which spurred consumer spending and construction (Kasrield, 1998). The persistent warm temperatures of the spring helped retailers in both April and May, with record sales gains noted in May for the fifth straight month (Wisely, 1998). For example, Kmart reported May clothing sales up 6.9 percent and attributed this to El Niño's affect in producing unseasonably warm weather (*Detroit News*, June 5, 1998). The record seasonal sales of goods and homes brought sizable added incomes to retailers, realtors, and homeowners, and summation of the various reported gains produced a national total estimated at $5.6 billion above normal expenditures. It is worth noting that these marked increases in retail and home sales occurred when the nation's economy was quite robust.

The early fears about bad weather brought economic predictions of instability in the commodity markets (Wilson, 1997). As a result, many brokers did a brisk business during the fall and winter of 1997 (see Figure 6-7). Economists reported that the lack of Atlantic hurricanes and attendant losses had been a major boon to insurers, further affecting investors, who increased their purchases of insurance stocks (Stead and Thomason, 1998). The lack of hurricane losses, as well as the drop in losses from spring snowmelt floods, benefited the federal government, which normally faces large relief costs related to hurricane and flood damages (NRC, 1999).

As a result of severe floods earlier in 1997, California was in the process of undertaking major mitigative activities when the El Niño–based climate predictions calling for a stormy and wet 1997–1998 cold season were issued in June 1997. The state spent an additional $7.5 million to aid in preparedness and to alert the public (Andrews, 1998). As a result of prompting by FEMA and the state, several California communities made special efforts and used their funds (e.g., Oakland spent $3 million) to reduce local flood and shoreline damages (FEMA, March 4, 1998). Flooding was still a major problem, however, as illustrated in Figure 6-8. No cost figures exist to measure the benefits of the mitigative activities undertaken in California, but the state suffered fewer losses in the winter of 1997–1998 (a total of $1.1 billion), than in the equally severe El Niño winter of 1982–1983 (approximately $2 billion after adjusting for inflation to 1998 dol-

---

The El Niño weather pattern is here, and it's bigger than ever. No one is sure, however, where it will do the most damage, if any. The only thing we can compare it to is the last major El Niño in 1983 which triggered $9.00 a bushel soybeans.

Since damage can occur anywhere in the world, a general strategy is to buy food commodities such as sugar, cocoa, coffee, corn and soybeans and maintain long positions throughout 1998.

Since timing the major impact of El Niño is too difficult, the best strategy is to maintain these long positions in the form of call options. Look to buy call options in the coming months which allow you to take advantage of the El Niño situation. For example, El Niño is expected to have its strong influence on the U.S. soybean crop in early summer. The best option to trade to take advantage of this event would be the November soybean options.

Finally, even if El Niño does not result in great crop damage, commodity markets are still likely to experience strong volatility due to "El Niño nervousness."

FIGURE 6-7. Excerpts from an advertisement of a brokerage firm, issued in October 1997.

FIGURE 6-8.　Workers busy filling and piling sandbags at Colusa, California, to keep rising waters from flooding nearby farm fields during mid-February 1998. (Courtesy Robert A. Eplett, California Office of Emergency Services)

lars). California roofing companies and home repair companies had major increases in business beginning in September 1997 and reported $125 million in added income as a result of El Niño–related mitigation concerns (Department of U.S. Labor, March 7, 1998).

Western water systems should also have benefited, since the director of the U.S. Geological Survey reported to Congress in October 1997 that government water managers in the Survey and Bureau of Reclamation had been instructed to plan their management strategies using the El Niño–based long-range forecasts calling for heavy precipitation (Shaefer, 1997). Sampling of several water managers (see chapter 5) revealed that many had used the predictions for planning purposes. The U.S. Secretary of Commerce acclaimed the correctness of NOAA's El Niño predictions, reflecting on the national benefits resulting from their use (Daley, 1997). NOAA Administrator Baker (1997), in presenting testimony before Congress, claimed that the value to U.S. agriculture of using El Niño predictions would be $275 million.

The net effect on the nation's economy from these varied benefits was detectable. For example, the Federal Reserve Board announced in February that the warm January had caused a 4 percent drop in production at the nation's electric and gas utilities, ending a run of months with production increases, which economists had expected to be + 0.3% in January (Federal Reserve Board, 1998).

El Niño's net influence on the weather and the Asian financial crisis combined in February to eliminate inflation in the prices paid by wholesalers as food processors and manufacturers charged wholesalers 0.1 percent less than in January 1998 for finished goods (Department of U.S. Labor, March 14, 1998). Inflation was held to zero during January–March, for the first time in ten years, and the Consumer Price Index went unchanged because of the falling energy prices (U.S. Department of Commerce, April 1998). The Gross Domestic Product rose at a torrid rate of 4.2 percent during the first quarter of 1998, as compared to the 3.4 percent expected (U.S. Department of Commerce, May 1998). Some of the gains are illustrated by the headlines in Figure 6-9.

Other outcomes partially attributed to El Niño–created conditions are difficult to quantify. For example, gasoline prices in the United States reached record lows in early March 1998, and oil experts indicated that part of the cause was the warmer-than-usual winter in the United States, which greatly reduced demand for oil, and part the Asian financial crisis and the bickering over sale quotas by the world's oil producers (*USA Today*, March 10, 1998). These lower gas prices, which averaged $0.25 per gallon below pre–El Niño costs, continued through 1998, represented an enormous saving to drivers. With 260 million autos and trucks operating and using an estimated ten gallons of fuel per week, the savings for just the March–May 1998 period amounted to $7.5 billion. But, how much of this can be attributed to the El Niño–caused warm winter? Even if only a small amount of the savings was attributable to El Niño, then very large benefits accrued across the nation, with gasoline prices remaining low throughout 1998. The head of the Energy Information Administration stated that the decrease in gasoline prices was largely a result of the winter's warmth (Stampler, 1998), which suggests that some of the savings benefits should be counted, but may were not included in the following list of gains.

Some of the national economic gains due to weather conditions attributed to El Niño can be estimated, and some are illustrated in Figure 6-9. The estimated gains are as follows:

- Reduced heating costs saved $6.7 billion.
- Increased sales of merchandise, homes, and other goods produced $5.6 billion.
- Reductions in costs for street and highway removal of ice and snow were estimated at $350 to $400 million.
- Reductions in normal losses because of the lack of snowmelt floods and the absence of Atlantic hurricanes was $6.9 billion.
- Income from increased construction and related employment was $450 to $500 million.
- Reduced costs to airline and trucking industry reached $160 to $175 million.

FIGURE 6-9. Headlines tell the story of the benefits and economic gains resulting from the El Niño–generated weather conditions from October 1997 to July 1998.

A NATIONAL ACCOUNTING

Various sources of data were employed to derive estimates of the many major financial losses and gains resulting from the weather conditions attributed to El Niño 97–98, as well as the effects on human lives. However, some impacts and most delayed effects have not been accurately measured or estimated at this writing. For example, the environmental impacts resulting from the weather conditions created by El Niño 97–98 are not well defined. Some were negative and others positive. For instance, the enhanced precipitation in the arid West certainly improved water supplies. The western rains filled reservoirs and also reduced energy costs, since hydroelectric plants could operate at full capacity.

Many envisioned that the mild winter would lead to increased insect pests in 1998; although the warm winter and spring led to early outbreaks of some pests, no unusually large summer outbreaks were reported. The many impacts resulting from use of the El Niño–based climate predictions for fall, winter, and early spring conditions represent another group of positive but largely unmeasured outcomes. A good example are the benefits derived from the widespread mitigation activities in California. The lower losses in California compared to those from similar El Niño events was sizable—$2 billion in the 1982–1983 event (adjusted to 1998 dollars) versus $1.1 billion in 1997–1998—suggesting the extensive mitigation activities in 1997–1998 (which cost an estimated $165 million) were beneficial. Utilities that used the winter forecasts and waited to purchase their natural gas supplies on the spot market during the winter as prices fell rapidly also reaped sizable benefits for their customers. One Iowa-based utility saved $39 million from the use of the predictions (Waetke, 1998), and two utilities in Michigan reported forecast-based savings, one of $48 million and another of $147 million (Bishop, 1998).

Most impacts identified fall into the category of direct losses and benefits and do not include many of the delayed effects that occurred well after El Niño ended. An example of the delayed effects are the Florida fires in June 1998, which damaged orange groves. This damage reportedly led to a 20 percent increase in orange juice prices in October 1998 (Cornell, 1998). Another good example of delayed impacts relates to the low use of natural gas in the northern states during the warm winter of 1997–1998. This led utilities to buy natural gas at low prices and to fill their gas storage fields, resulting in abundant supplies for the 1998–1999 winter. As a result, natural gas prices in 1998–1999 were 20 percent to 30 percent lower than in normal winters (*Wall Street Journal*, December 7, 1998). Delayed, hard-to-measure impacts include the repair and rebuilding of storm- and flood-damaged structures. These delayed effects will eventually bring profits to the construction industry. Some events described as El Niño–related were probably not. For example, the drought in Texas during summer 1998, which began as a dry spring in 1998, was an outcome in direct contrast to the El Niño–based forecasts for wet conditions in Texas and was not attributed to El Niño–related conditions.

A summary of the national impacts, both losses and benefits, appears in Table 6-1, which reveals that the benefits outweighed the losses, both in human lives and in the financial outcomes. Michaels (1998) made an early estimate that showed national benefits of $15 billion and losses of $2 billion. (Note that the dollar values in Table 6-1 are based on estimates that may be in error by as much as 30 percent. Thus, the losses could be somewhere between $3 and $6 billion, and the gains between $14 and $25 billion).

The sizable and unexpected benefits from El Niño conditions were widely reported by the press. One typical news article stated, "Effects of El Niño are mostly a positive outcome." (*Reuters*, March 17, 1998). This net positive outcome led

TABLE 6–1. The national tally of impacts resulting from El Niño–related weather conditions during 1997–1998.

Losses	Gains
Human lives = 189 Economic losses and costs = $4.2 billion to $4.5 billion	Human lives saved = ~850 Economic gains = $19.6 billion to $19.9 billion

another assessor of the impacts to contrast this outcome against the climatologists' predictions in 1997, which called for major losses (*Cincinnati Inquirer*, March 26, 1998). This news assessment concluded that the erroneous prediction of "bad impacts" raised major doubts about scientists' predictions of the negative outcomes likely to result from global warming.

IMPACTS IN THE MIDWEST

This study identified for the Midwest the myriad impacts that resulted from the uses of El Niño 1997–1998 climate predictions and from the actual El Niño–driven weather conditions (Changnon et al., 1999). Where possible, information on the magnitude of the impacts is presented. A regional analysis, performed in much greater detail than the national analysis, helps reveal the varied dimensions of the myriad impacts of El Niño's weather.

Major Impacts Created by El Niño in the Midwest

Assessment of the impacts of El Niño on the Midwest requires an understanding of the temporal history of El Niño 97–98. The Climate Prediction Center began issuing official predictions late in June 1997 about the development of a major warm phase (El Niño) in the tropical Pacific (CPC, June 1997). By August, the official CPC predictions were calling for mild fall and winter seasons, with below-normal precipitation in the Midwest (see Figure 5-4). All other experimental forecasts by other scientists and institutions agreed. The Midwestern Climate Center (MCC) issued, in September, a climate-based outlook that called for much-below-normal snowfall for the Midwest during the winter of 1997–1998. An MCC outlook for decreased cold-season storminess on the Great Lakes, issued in November, was countered by scientists at the University of Michigan, who predicted increased lake storms. These differing predictions received wide coverage in the Midwestern news media (see chapter 4).

By late September, most Midwestern residents and decision makers in weather-sensitive industries were aware of the predictions for the fall and winter and were in a position to make decisions relating to the forecasted winter conditions. The first series of impacts that occurred during the fall of 1997 largely related to decisions to use or to ignore the El Niño–based forecasts for a mild, dry, low snowfall winter across the Midwest. Many were not impressed with the predictions, since the fall season weather did not reflect any well-defined ENSO effects (CPC, November 1997).

The ensuing mild, relatively dry, and almost snowless winter (December 1997–February 1998) revealed the official forecasts to be accurate (see Figure 5-5). NOAA announced, on March 9, 1998, that the winter had been the second warmest on record and that February 1998 was the warmest ever in the Midwest. It was during this season that the major impacts from the El Niño–driven weather conditions occurred across the Midwest. These impacts transcended most human and economic activities in the Midwest. As will be shown, the impacts were realized in numerous sectors, including human health and safety, finance, energy production and use, societal activities, agriculture, business, recreational activities, and the environment.

The third series of Midwestern impacts and responses was related to the climate predictions issued during January–April, which forecast the late spring and the 1998 growing season (May–September) conditions. These were of great interest to Midwestern agricultural interests and to the commodity-investment markets. The various climate predictions for summer 1998, unlike those for the winter, were not in agreement. Some called for a mild and wet summer, whereas others called for high temperatures and drought. The spring season predictions, issued in February, called for a continuation of the winterlike departures, with predictions for warmer and drier-than-normal conditions for the Midwest. The spring weather in the Midwest had above-normal temperatures but also above-normal precipitation, instead of drier-than-normal conditions. Furthermore, March brought a two-week period of cold and snowy weather, with two major winter storms temporarily interrupting the mild cold-season conditions.

Considerable information on the impacts of El Niño was extracted from articles published in eight major Midwestern newspapers that contained any El Niño references. The number of newspaper articles assessed for El Niño information totaled 502. More than 200 Internet issuances pertaining to El Niño and the Midwest, as found during the July 1997–May 1998 period, were also assessed for impact information, and considerable information was obtained from government agencies and from the staffs of various businesses.

Impacts and Responses to the Climate Predictions for Fall and Winter

Predictions of wet and stormy cold season conditions from California east through the Gulf States to Florida quickly got the attention of various persons interested in agriculture. By mid-September 1997, a commodities group in Chicago predicted food prices would rise (*Cleveland Plain Dealer*, September 18, 1997), a prediction echoed by a leading economist a few days later (*Detroit News*, September 22, 1997). A private-sector meteorologist in the Midwest predicted the types of crops (produce and fruit) that would be hurt by El Niño weather and added that Midwestern prices would soon rise (*Cincinnati Inquirer*, October 3, 1997). In a more cautious vein, and after El Niño's effects on fall weather conditions had been demonstrated, the U.S. Department of Agriculture (USDA) predicted, in mid-December, that higher fruit and vegetable prices would exist in the first six months of 1998 because of bad weather expected in California, Arizona, Texas, and Mexico (*Chicago Tribune*, December 17, 1997). Interestingly, events in January and succeeding months would prove that these agricultural impact projections were essentially correct.

Several companies began increasing their stocks and supplies in October in anticipation of buyers' reactions to the El Niño–based predictions of increases in damaging storms (*Minneapolis Star Tribune*, October 27, 1997). Big hardware and lumber chains began increasing their stocks, the insurance industry reported increased sales of policies and additions to existing policies, and commodity traders, who were betting on the forecasts for dry conditions in the Corn Belt and elsewhere, had increased purchases from investors. Some homeowners, fearing the worst, got home repairs, and this benefited the construction industry.

Varying responses to the El Niño–based seasonal predictions were issued by Midwestern energy experts. One expert predicted that the projected warm winter would have little effect on the use of natural gas in the Midwest (*Des Moines Register*, October 26, 1997). Another predicted a major drop in the prices for natural gas (*Chicago Tribune*, December 5, 1997), a correct prediction. More conflicting predictions were issued concerning the incidence of damaging winter storms on the Great Lakes. One climatologist predicted fewer storms than normal as a result of El Niño's effect on storm tracks (University of Illinois, 1997), whereas other scientists predicted bad lake storms and much shore erosion (University of Michigan, 1997). The ensuing winter had below-normal lake storms and minimal erosion, validating the Illinois prediction.

Experts in the travel and tourist industry reacted to the predictions with forecasts of bad business for winter sports and skiing in the Midwest (*Des Moines Register*, November 16, 1997; *Detroit News*, December 10, 1997), a generally correct prediction. Financial specialists foresaw instability in Midwestern commodity markets (*Chicago Tribune*, September 25, 1997), a condition that did not develop. Economists predicted that the expected winter weather would have mixed effects on Midwestern businesses, a cautious and correct outlook (*Chicago Tribune*, January 7, 1998).

The reactions of various groups and the public to the El Niño–based predictions for the mild winter weather were interesting. A group of 100 Iowans was polled late in October, and the majority doubted the winter forecast (*Des Moines Register*, November 3, 1997). A mid-November assessment of public attitudes across the Midwest indicated that the majority of the general public was confused by the El Niño forecasts. Experts interviewed at three Midwestern natural gas supply companies questioned the accuracy of the predictions and indicated they were all going to ignore predictions of a mild winter; it was going to be business as usual (*Cleveland Plain Dealer*, November 9, 1997). Managers responsible for clearing highways and streets in several Midwestern cities also expressed doubt in the predictions; all those sampled had decided to "play it safe" and assume the worst in planning for the number of road crews and the amount of salt to have available (*Chicago Tribune*, January 18, 1998). Thus, there was widespread disbelief in the seasonal predictions, and only a few officials responded and acted to follow them. As will be shown, this was unfortunate, since those who did respond to the predictions realized major gains.

Impacts from the Fall-Winter-Spring Weather Conditions

The various impacts associated with the El Niño–derived weather conditions in the Midwest from October through May 1998 were classified by sector, beginning with the societal impacts, including effects on behavior and health. We also assessed the impact of the weather on business and industry, energy use and production, the recreation industry, the construction industry, agriculture, the environment, and, finally, government agencies.

Societal Impacts. Societal impacts include behavioral effects on human health. Midwesterners quickly adopted, as did the entire nation, the concept that El Niño was a good excuse for everything bad, including personal failures (*Detroit News*, October 30, 1997; *Minneapolis Star Tribune*, October 22, 1997). It was noted that street and highway workers, because they performed fewer road clearance activities, were doing less work, had more rest, and experienced less stress than in normal winters (*Chicago Tribune*, January 7, 1998).

Public sampling done in Ohio during late January (*Cleveland Plain Dealer*, February 2, 1998), in Michigan in early February (*Detroit News*, February 6, 1998), and in Chicago during mid February (*Chicago Tribune*, February 16, 1998) revealed most people were quite happy with the nice winter. They reported the weather had allowed them to be outdoors more, as illustrated in Figure 6-10, and had substantially lowered their heating costs. Only a few complained about the lack of snow-related sports.

The favorable conditions led to many fewer winter highway accidents and the absence of excessively low temperatures reduced the number of deaths in the Midwest by 230 persons (*Detroit News*, March 20, 1998). There were notably fewer winter injuries, such as falls on slippery surfaces, and the reduction in injuries was most notable among the elderly (*Detroit News*, March 20, 1998). Doctors in several hospitals reported seeing 22 percent fewer heart attacks overall. The homeless were seen as beneficiaries of the mild winter (*Cleveland Plain*

FIGURE 6-10.   Shoppers fill North Michigan Avenue in Chicago on a mild afternoon in February 1998. The mild Midwestern winter led to extensive shopping, setting all-time records for retail sales in January, February, and March.

Dealer, March 1, 1998), and emergency highway related calls were down by 11 percent in Michigan during December–February (*Detroit News*, March 20, 1998). In all, the number of lives lost during the winter dropped by between 250 and 350 people in the Midwest.

In general, it was healthier and safer to be outdoors and to travel during the winter of 1997–1998. However, deaths related to El Niño's weather did occur. Three persons riding on ski mobiles over northern Michigan lakes, which are normally covered by thick ice, fell through the thinner ice in February 1998 and drowned (*Detroit News*, February 15, 1998). A huge storm system that developed over the South and was attributed to El Niño produced eighteen inches of snow in Kentucky and Ohio and floods elsewhere along the East Coast, leading to four deaths in the Midwest. A tornado in Minnesota in March killed two, and a National Weather Service (NWS) spokesman attributed the storm to El Niño (*Cleveland Plain Dealer*, March 31, 1998). There were no spring floods in the Midwest, events that normally cause considerable regional damage and a few deaths. The huge flood in April 1997 on the Red River of the North had devastated parts of Minnesota and North Dakota, and the nice winter weather of 1997–1998 was considered a huge blessing to morale in that area, as people were able to rebuild their damaged homes faster than had been expected (*Chicago Tribune*, March 23, 1998).

Effects on Business and Industry. The unexpected and persistent good winter weather led the buying public to circulate and behave in many interesting and unusual ways (*Minneapolis Star Tribune*, November 17, 1997). The region experienced a major increase in housing starts in December and January (*Chicago Tribune*, January 7, 1998). Realtors reported that the December–March sales of homes in the Midwest were up by 25.5 percent, reaching the highest level in four years (*Chicago Tribune*, March 30, 1998) and representing added income to realtors and homeowners of $0.6 billion (National Association of Realtors, 1998). National sales of homes in January alone reached a record high at 4.4 million units (*Chicago Tribune*, April 5, 1998), and these record sales were attributed to the mild and snowless weather created by El Niño.

Restaurants reported record crowds as people went out more than they did in normal cold and snowy winters (*Detroit News*, January 8, 1998), and retail sales skyrocketed for the same reason. In early January, major Midwestern department store chains such as Dayton-Hudson, Nordstroms, and Neiman Marcus, as well as major shopping centers, reported that winter season retail sales were up by 5 percent to 15 percent (*Chicago Tribune*, January 7, 1998). Retail sales, including sales of clothing and products for yard work, continued to boom during February. Wal-mart reported a 9.6 percent increase in sales over the previous winter, Sears's sales were up 4.7 percent, and Kmart's increase was 5.6 percent (*Chicago Tribune*, March 5, 1998). Such increases represented additional incomes of over $1.5 billion to these businesses, as well as to other retailers. The nation's retail index was up by 5.4 percent in February after a gain of 5.1 percent in January, and a group of major Midwestern department stores reported winter sales were up by 7 percent (*Chicago Tribune*, March 6, 1998). These winter and spring outcomes across the Midwest produced a total benefit from increased merchandise and home sales estimated at $2.9 billion.

Transportation and shipping systems in the Midwest benefited widely from the unusual winter conditions. With less bad weather, trucks and railroads experienced fewer delays and accidents. Airlines operating at Midwestern airports also had many fewer delays and experienced increased rider ship. The warm conditions reduced ice cover on the Great Lakes and allowed shippers to haul more tonnage than in any other year since 1982 (*Great Lakes Update*, 1998). The estimates of the resulting cost savings and increased profits from these conditions ranged between $80 million and $100 million.

Not all Midwestern businesses benefited from the winter weather conditions. Companies that manufactured or sold snow equipment (e.g., snow mobiles, snow plows, shovels) were hurt by low sales. Snowmobile sales were down 35 percent (*Detroit News*, March 20, 1998), a major snowplow manufacturing company in Ohio laid off half its staff (*Cleveland Plain Dealer*, March 1, 1998), and another Ohio-based manufacturer went broke. Sales of salt were down 50 percent, and sales of winter clothes and fur coats went down 15 percent (*Detroit News*, March 20, 1998). Midwestern salt providers reported losses of $68 million in sales. Urban taxis reported fewer riders (*Chicago Tribune*, January 7, 1998), and manufacturers and outlets for furnaces reported sales down by 20 percent. Towing companies reported business down by 40 percent. Natural gas providers had reduced sales because of the above-normal temperatures, with winter sales down by 15 percent. Companies that had made plans based on the predicted winter conditions benefited by buying gas supplies at low prices during the winter. Others that had contracted for gas in the early fall, when the price was still high, were able to pass on the higher price paid to the consumer by not lowering rates as much during the winter. Several of the economic impacts of El Niño–based weather are illustrated by the headlines in Figure 6-11.

Energy Use and Production. The higher temperatures, many at near-record levels for winter, greatly decreased the use of natural gas and heating oil in the Midwest. Use was down by 15 percent to 30 percent, depending on the area. By early December, gas futures, after reaching a high in October, were down by 36 percent (*Chicago Tribune*, December 5, 1997). This was reflected by the amount of natural gas in storage. Regional storage in early December was 88 percent of capacity, compared to 77 percent a year before. In the Chicago area, gas prices

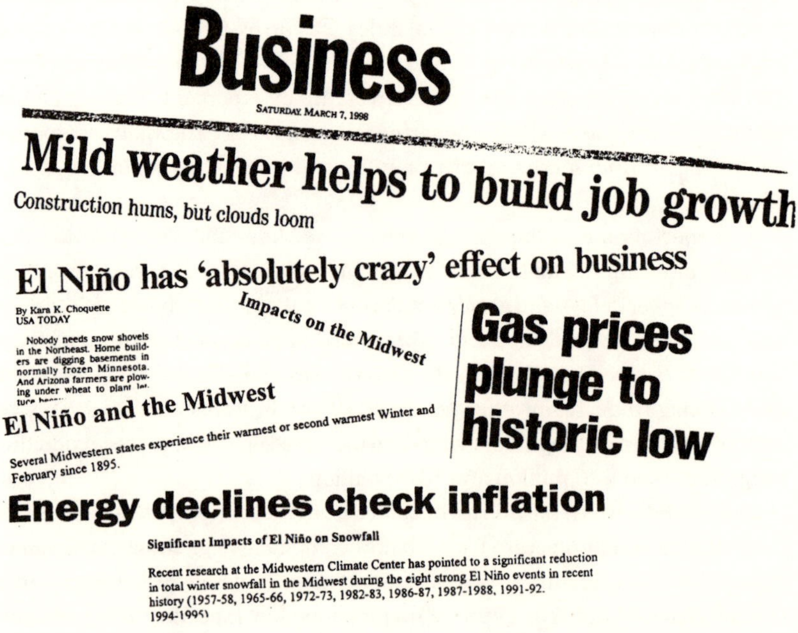

Business
SATURDAY, MARCH 7, 1998

Mild weather helps to build job growth
Construction hums, but clouds loom

El Niño has 'absolutely crazy' effect on business

By Kara K. Choquette
USA TODAY

Impacts on the Midwest

Nobody needs snow shovels in the Northeast. Home builders are digging basements in normally frozen Minnesota. And Arizona farmers are plowing under wheat to plant let-tuce here...

El Niño and the Midwest

Several Midwestern states experience their warmest or second warmest Winter and February since 1895.

Gas prices plunge to historic low

Energy declines check inflation

Significant Impacts of El Niño on Snowfall

Recent research at the Midwestern Climate Center has pointed to a significant reduction in total winter snowfall in the Midwest during the eight strong El Niño events in recent history (1957-58, 1965-66, 1972-73, 1982-83, 1986-87, 1987-1988, 1991-92, 1994-1995)

FIGURE 6-11. The mild, dry, and largely snow-free winter of 1997–1998 in the Midwest brought many benefits, as illustrated by these headlines from regional newspapers.

fell, on average, by 3.3 percent in December. Gas prices to consumers fluctuated, some remaining relatively high and others cascading, depending on whether their suppliers had contracted early when prices were up or waited and were able to buy gas at the ever-lowering prices. Some of the gas utilities that used the long-range climate predictions and waited to buy on the spot market during the winter gambled further and used the gas they had in storage early in the winter, allowing them to wait to buy as the price fell ever lower (*Des Moines Register*, February 23, 1998).

By late February, energy bills for the winter in Ohio, Iowa, and Illinois had decreased 16 percent, and the average person had saved $50 (*Cleveland Plain Dealer*, March 1, 1998; Des *Moines Register*, February 23, 1998). Figure 6-6 illustrates the magnitude of savings at one Midwestern home. One Iowa utility reported that its 610,000 customers had realized savings of $39 million. By late February, savings from lower heating costs in the nine-state Midwest had reached $4.0 billion, and more savings lay ahead with the continuing above-normal temperatures of March and April. Utilities in lower Michigan announced that gas usage in March was down 18 percent, the average savings per person was $21, and the region's one-month savings were $125 million (*Detroit News*, March 12, 1998).

As the demand for heating fuels fell, global energy prices also declined, falling 2.4 percent in January and 2.2 percent in February, and this helped bring down the price of gasoline to $1.10 per gallon, the Midwestern average price in early March (*Chicago Tribune*, March 20, 1998). This was down 25 cents per gallon from the average in prior months. The only energy-related problem reported was a major power outage in southern Ohio and northern Kentucky that resulted from a major snowstorm early in February (*Des Moines Register*, February 7, 1998). By April 30, the savings in energy costs in the Midwest were estimated at $4.7 billion. The variety of benefits realized in the Midwest are illustrated in Figure 6-11.

Impacts on the Recreation Industry.　The effects of the mild winter on recreation in the Midwest were mixed, with some winners and some losers. The major losses hit the ski industry in Michigan, Ohio, and Wisconsin. Overall, seasonal sales at Michigan's ski resorts were down 50 percent, a loss of $25 million, but those resorts that had, or added, snowmaking equipment (some relying on the mild winter forecasts) reported a good financial season (*Detroit News*, December 10, 1997). A major ski resort in northern Ohio spent $2.5 million on snow-making equipment and reported that the benefit from the equipment exceeded its cost (*Cleveland Plain Dealer*, March 1, 1998). Several ski resorts had to close two to four weeks earlier than normal because they could no longer make enough artificial snow, given the unusual warmth and rainfall (*Detroit News*, March 20, 1998). Cross-country skiing was also hindered, as was ice skating (*Des Moines Register*, January 14, 1998). The thin-ice problem of the warm winter also curtailed ice fishing across the Midwest. However, boat fishing on Lake Erie in mid-winter was possible because there was no lake ice to contend with, and this was a big boon to fishermen and to the local boating industry (*Cleveland Plain Dealer*, February 13, 1998).

Many other positive impacts came in the recreation sector. Golf courses reported golfers were out in record numbers all through the winter. A Chicago course reported that seventy-five groups had played on January 5, a record for any day in any previous winter (*Chicago Tribune*, January 7, 1998). Chicagoans in February were enjoying their parks, walking along the lake, roller skating, and sailing on Lake Michigan, all activities not common in winter (*Chicago Tribune*, February 16, 1998). Attendance at city zoos set records (*Cleveland Plain Dealer*, March 1, 1998). Many skiers went to the West to ski, leading to increased skiing income in California and Colorado (*Cincinnati Inquirer*, March 22, 1998). Among the oddities reported was the fact that owners of classic cars were able to operate their vehicles in winter because the roads were not salted, and the mild temperatures allowed track athletes to begin practice earlier than usual and to improve their racing times (*Detroit News*, April 2, 1998). In sum, many people found different outlets for their winter recreational needs, but those most unhappy with the winter of 1997–1998 were skiers, ice skaters, and ice fishermen.

Effect on the Construction Industry.   The mild temperatures and low precipitation allowed building construction to proceed at a fast pace in the Midwest, as illustrated in Figure 6-11. Winter construction projects in Ohio and in Michigan were up 30 percent (*Cleveland Plain Dealer*, March 1, 1998; *Detroit News*, March 20, 1998). People not only started building houses, but many added on to existing structures. Cincinnati construction firms reported losing only one day to bad weather between December 1 and February 28, as compared to fourteen bad-weather days in a normal winter (*Cincinnati Inquirer*, March 20, 1998). The good weather also increased Midwestern construction employment by 4.7 percent, another benefit (*Cincinnati Inquirer*, April 4, 1998). The resulting economic benefits were estimated at between $250 million and $300 million in the Midwest.

Impacts on Agriculture.   Since the cold season is not the growing season, Midwestern agriculture did not experience major impacts from the winter weather of 1997–1998, but most effects reported were positive. Farming impacts during the winter included lower costs for heating farm buildings. Herds of cattle and sheep could be kept larger because feed was less expensive and animals could be outside a greater percentage of the time consuming grasses. Gardens got an early start (*Des Moines Register*, March 8a, 1998). Although many orchardists feared and then witnessed early budding of their fruit trees, no freezes came late to damage trees (*Farm Week*, April 6, 1998), and apple and peach crops in the Midwest had above-normal yields in 1998, leading to added profits of $85 million (*Champaign-Urbana News Gazette*, August 1, 1998). The sale of crop insurance in the spring rose 21 percent above average levels, representing farmers' fears for the 1998 growing season. However, no drought developed, and corn and soybean yields across most of the Midwest were near or slightly above average (*Farm Week*, January 1999). A major benefit of the warmer-than-normal spring was that it allowed many Midwestern farmers to plant their corn crops earlier than usual, thus averting the potential of serious plant stress during critical growth periods if high temperatures came in July. Early planting was also very beneficial in Illinois and Indiana because heavy rains in late April and May kept farmers from their fields until mid-June.

Environmental Impacts. The warm and dry winter had predictable environmental outcomes. Trees and bushes began budding earlier than normal (*Cleveland Plain Dealer*, March 1, 1998). Wisconsin authorities reported an increase in the deer population (*Chicago Tribune*, March 2, 1998). Since wild birds could access natural feeds during the winter, they consumed less human-supplied bird feed, leading to reduced sales of commercial feeds (*Cincinnati Inquirer*, March 4, 1998). The number of bald eagles was reduced because the eagles, which normally winter in Missouri, spent the winter at habitats farther north (*St. Louis Post-Dispatch*, March 9, 1998). Because some tree varieties put out buds early, they were vulnerable to damage when an early March snowstorm occurred, the only significant winter storm of the year across the Midwest (*Chicago Tribune*, March 10, 1998). Stream measurements taken throughout the region also revealed that water pollution levels in most streams and rivers were lower than normal (*Detroit News*, March 20, 1998). Stream flows were at or below average as a result of the drier winter but did not approach serious shortage levels in streams or reservoirs, and shallow ground water levels remained near normal in Illinois (Illinois State Water Survey, 1998). The levels of the Great Lakes declined by six inches to one foot as a result of the dry and warm cold season, but this was largely a benefit, since lake levels had been much above normal, with waves producing shoreline damages. Furthermore, the much-below-normal number of winter lake storms reduced damage to the shorelines (*Great Lakes Update*, 1998).

Impacts and Responses of Government. The principal impact of the mild and snow-free winter of 1997–1998 on city and state governments was reduced costs. These occurred because of the lack of snowstorms and ice storms, which greatly reduced the amount of salt needed and purchased and reduced overtime payments to road crews for snow removal. For example, the Ohio Department of Transportation used only one-third the amount of salt normally applied, saved 10,000 hours of overtime pay, and reported a total savings of $930,000 (*Cleveland Plain Dealer*, March 1, 1998). Counties across Michigan realized 40 percent cuts in overtime payments and reported an average winter cost reduction of $1 million per county, representing a statewide gain of $100 million; for example, Wayne County saved $200,000 in overtime pay and $2.4 million in salt purchases (*Detroit News*, March 20, 1998). Illinois collar counties around Chicago reported winter savings ranging from $100,000 to $250,000 each, and Chicago spent $8 million for snow removal, which was $4 million less than normal (*Chicago Tribune*, April 1, 1998). The Illinois Department of Transportation used only 50 percent of the normal amount of salt in the Chicago metropolitan area, a savings of $2 million. The metropolitan region's total savings amounted to $21 million. Use of such values to estimate the Midwestern benefits suggests a regional savings of approximately $250 million. The increased retail sales and home sales also added taxable income to the state and local coffers, and the lower use of energy cut heating costs for government buildings. The impacts of El Niño on government bodies were beneficial.

Impacts and Responses to Climate Predictions for the Summer of 1998

Analyses of the official and nonofficial experimental predictions issued by various groups from midwinter on about the spring and summer climate conditions, which are described in chapters 2 and 4, revealed a wide variety of predicted outcomes, particularly for the late spring and summer—the growing season of 1998. The official forecasts of sea surface temperatures (SST) in the tropical Pacific issued in February (CPC, February 1998) indicated "substantial disagreement" among the various forecast models for the SST conditions expected by mid-1998. Some of the predictions called for a continuation of the El Niño–type conditions into June, whereas others foresaw a rapid development of La Niña conditions, resulting in drought-like conditions in the Midwest. The point is that the various scientific models used to generate the seasonal predictions did not agree and provided a wide range of forecast outcomes (Kirtman, 1998). The El Niño–based outlooks for rainfall across the Midwest in the early spring of 1998 (Figure 6-12) indicated a strong likelihood of dry conditions.

These uncertainties were translated into various predictions of the effects on agriculture, a major concern in the Midwest. Early in February, an Iowa climatologist forecast normal corn and soybean yields for Iowa (a correct forecast), but other weather experts in Iowa disagreed (*Des Moines Register*, February 19, 1998). Five days later, a USDA economist in Washington predicted "bumper" corn and soybean yields for the United States in 1998 (*Chicago Tribune*, February 23, 1998), a forecast that was incorrect for the Midwest. Five days later, a private-sector meteorologist in Chicago stated that there was 57 percent chance of bad growing season weather (due to high temperatures) in the Midwest and further projected that future corn and soybean prices would soar (*Chicago Tribune*, March 1, 1998). Both of these predictions were incorrect. The USDA regional crop insurance office in the Midwest recommended buying coverage for what was expected to be a growing season marked by bad weather in 1998 (*Des Moines Register*, March 8b, 1998).

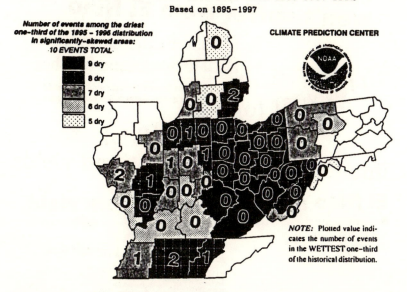

Significantly-Skewed El Niño Precipitation Distributions — *January - March*
1915 1919 1941 1958 1966 1969 1973 1983 1987 1992
Based on 1895–1997

Number of events among the driest
one-third of the 1895 - 1996 distribution
in significantly-skewed areas:
·10 EVENTS TOTAL·

9 dry
8 dry
7 dry
6 dry
5 dry

CLIMATE PREDICTION CENTER

NOTE: Plotted value indicates the number of events
in the WETTEST one-third
of the historical distribution.

FIGURE 6-12.   The likely distribution of precipitation during January–March in the
Midwest. The values shown are based on conditions in ten past periods of strong El
Niño events. This statistical distribution suggests extremely low precipitation during
such periods. (Climate Prediction Center)

A sampling of various crop-weather experts early in March revealed a wide
range of predicted Midwestern crop outcomes, from poor to great, and one
weather expert wisely observed that they should all be ignored since there was
no skill involved in the forecasts (*Farm Week*, March 6, 1998). An official forecaster for the NWS predicted a normal summer in the Midwest, a correct prediction (*Minneapolis Star Tribune*, March 18, 1998). An Iowa climatologist predicted a cold spring with above-average snowfall (later shown to be incorrect)
and also predicted that July and August would be hot and dry (*Des Moines
Register*, March 20, 1998), another outcome that did not occur. The media
noted that the different climate conditions being forecast raised serious questions about the ability of atmospheric scientists to correctly predict either the
end or the beginning of El Niño events (*Cincinnati Inquirer*, March 26, 1998).
NWS hydrologists issued a spring flood outlook (Richards et al., 1998), and there
were positive media responses to the predictions, which (correctly) called for a
lower incidence of floods as a result of the dry, snow-free winter (*St.Louis Post-
Dispatch*, March 25, 1998). The wide diversity of predicted outcomes for the 1998
growing season made the headlines, as shown in Figure 6-13.

Some medical experts predicted a bad warm season for those with allergies
because of the predicted warm spring, earlier budding, and lengthened growing
season (*Detroit News*, February 23, 1998; *Minneapolis Star Tribune*, March 29,

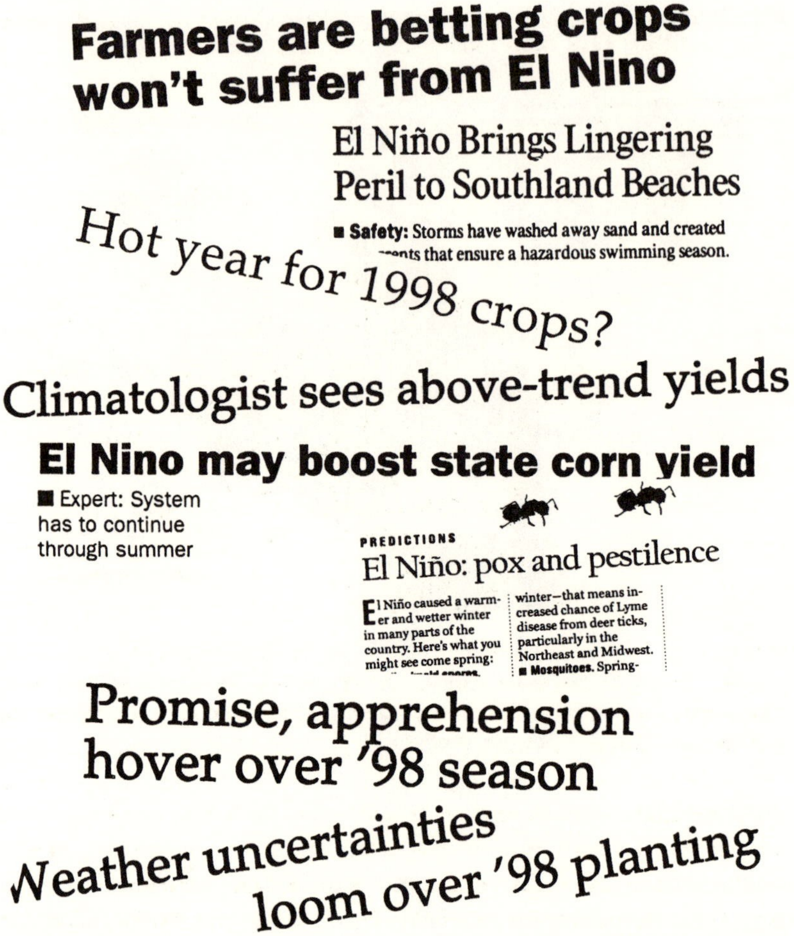

Farmers are betting crops won't suffer from El Nino

El Niño Brings Lingering Peril to Southland Beaches

■ **Safety:** Storms have washed away sand and created ~~nts that ensure a hazardous swimming season.

Hot year for 1998 crops?

Climatologist sees above-trend yields

El Nino may boost state corn yield

■ Expert: System has to continue through summer

PREDICTIONS

El Niño: pox and pestilence

El Niño caused a warm-
er and wetter winter
in many parts of the
country. Here's what you
might see come spring:

winter—that means in-
creased chance of Lyme
disease from deer ticks,
particularly in the
Northeast and Midwest.
■ **Mosquitoes.** Spring-

Promise, apprehension hover over '98 season

Weather uncertainties loom over '98 planting

FIGURE 6-13.   The climate forecasts issued by various groups during the winter of 1997–1998 for the weather conditions in the 1998 growing season differed, ranging from very good to very bad outcomes, and these created major uncertainties among Midwest agricultural decision makers.

1998; *Time*, March 23, 1998). Other medical experts predicted more asthma problems for 1998 (*Detroit News*, March 20, 1998). Statistics are not available to validate or refute these predictions.

Predictions were also issued about the effect of the warm winter and the predicted warm spring on insect pests. They predicted more mosquitos and other pests in the spring and summer (*Cincinnati Inquirer*, January 7, 1998). Others concluded that the weather predictions would improve Midwestern fishing (*St. Louis Post-Dispatch*, February 22, 1998). The pest predictions did prove correct for some parts of the Midwest where the spring had been warm and wet (Michigan, Illinois, Indiana, and Ohio). Outbreaks of flies, termites, and mosquitos began two to three weeks earlier than usual, and there were more flies than normal (*Detroit News*, June 23, 1998). However, in the western portions of the Midwest, where the spring had been warm but dry (as the El Niño forecasts had predicted; see figure 5-4), the population of mosquitos and other pests was at normal levels (*Minneapolis Star Tribune*, May 29, 1998).

SUMMARY

The mild, dry, and largely snow-free winter across the Midwest created an interesting array of impacts, and most were beneficial. The recreation industry had winners and losers, but, in general, the public, many businesses, and federal, state, and local government agencies were major winners. Some benefited by using the seasonal weather predictions to alter their normal operational plans, and most Midwesterners benefited in terms of health as well as financially from the mild winter and spring weather conditions.

The population of the nine-state Midwest is 60 million people, and it is safe to conclude from numerous samplings of public attitudes that most residents were pleased with the mild, almost snow-free winter. It was safer and healthier than a normal winter. There was much less illness and many fewer deaths than occur in a normal winter as a result of cold temperatures and accidents caused by dangerous travel conditions. Medical and travel experts concluded that the winter conditions led to the saving of between 350 and 380 lives. Many residents of the Midwest altered their recreation plans, and many used the mild weather to fish, play golf, and hike.

Those benefiting in business included commodity brokers, who profited from increased investments by buyers who feared El Niño–driven damages to foods grown in the United States and elsewhere. Insurers sold more policies and paid out fewer claims than normal. The construction industry enjoyed more favorable work conditions, and the region's income for construction was up $300 million over income in normal winters. Employment in retail businesses and construction was up by 4 to 5 percent. Home sales soared 25 percent over normal levels higher for December 1997–March 1998, and this additional income represented a gain of $0.6 billion.

The mild weather with below-normal rainfall and snow made it pleasant to be outdoors, and this led to major increases in retail sales of clothing, furniture, and other goods. Sales were $2.9 billion over the average December–March sales. Restaurants reported increased business, but no estimates of the increased income is available. The near-record warmth in the Midwest reduced heating costs, creating consumer savings amounting to $4.7 billion. The local and state government agencies saved $250 million thanks to reduced costs related to lower snow removal and other road-related costs. All forms of transportation and shipping by land, water, and air benefited from the nearly storm-free conditions, leading to savings and/or extra profits that reached $120 million.

In sum, the known financial benefits traceable to the unseasonal winter-early spring weather in the Midwest amounted to $8.85 billion. There may be additional but unknown income in some sectors. Little is known about the environmental impacts, but no significant environmental effects have been reported.

Those who lost out as a result of the El Niño–driven weather conditions in the Midwest included those who failed to use the predictions of a mild winter and those who were hurt by the untypical weather conditions. Several natural gas providers failed to heed the long-range weather predictions and bought gas early in the fall at high prices. These higher costs were largely transferred to their customers, who failed to realize all the benefits that later, lower-cost purchases would have produced. From a health standpoint, the unusual winter weather conditions resulted in nine deaths in the Midwest. Certain businesses suffered, including manufactures and sellers of snowmobiles ($10 million in lost sales) and manufacturers of snow removal equipment (lost $22 million). Private snow removal companies lost $26 million in expected income, towing businesses lost $16 million, and sellers of salt reported decreased sales and losses of $68 million. Business at Midwestern ski resorts was down 50 percent and losses were $120 million. Furnace manufacturers and heating equipment sales companies reported losses of $300 million. Retail sales of winter clothing, including fur coats, was down 15 percent, and this represented losses of $180 million. The total estimated losses in the Midwest amounted to $750 million. Thus, losses were approximately 10 percent of the total benefits.

In general, the climate predictions calling for widely varying weather conditions during summer 1998 and the growing season left decision makers with no guidance and, often, feeling confused. Hence, their responses to the predictions were negligible.

REFERENCES

Adams, C. R. 1997. Impacts of temperature extremes. *Proceedings of the Workshop on the Societal and Economic Impacts of Weather.* Boulder, CO: NCAR, 11–16.

Baker, D. J. May 15, 1997. *Testimony before Subcommittee on Science, Technology & Space,* Committee on Commerce, Science & Transportation, U.S. Senate, Washington, DC.

Bishop, D. March 12, 1998. El Nino led to lower gas bills. *Detroit News,* p. B1.

Bunting, V. April 4, 1998. Winter damage from El Niño bad but not the worst. *Champaign-Urbana News Gazette,* p. 11.

Champaign-Urbana News Gazette. August 1, 1998. State's peaches prove plentiful, p. 11.

Changnon, S. A. 1996. Losers and winners: A summary of the flood's impacts. In *The Great Flood of 1993.* Boulder, CO: Westview Press, 276–299.

Changnon, S., Changnon, D., Fosse, E., Hoganson, R., Roth, R., and Totsch, J. 1997. Effects of recent extremes on the insurance industry: Major implications for the atmospheric sciences. *Bulletin Amer. Meteoro. Soc.,* 78, 425–435.

Changnon, S. A., Hilberg, S., and Kunkel, K. 1999. *El Niño 1997–98 in the Midwest: The Predictions, the Weather Conditions and Their Climatological Relevance, and the Impacts of the Weather on Society and Economy.* Contract Report. Champaign, IL: Illinois State Water Survey.

Chicago Tribune. September 25, 1997. El Niño already causing headaches: Unusually warm water in stretches of the Pacific worrying traders, p. 1.

———. December 5, 1997. El Niño deflates gas prices, futures fall faster than utility bills, p. 1.

———. December 17, 1997. Winter produce prices depend on weather, p. 6.

———. January 7, 1998. Stormy and sunny effects of the season's warming, p. 1.

———. January 18, 1998. Snow foe, p. 1.

———. February 16, 1998. Chicagoland springs to life as El Niño warms up winter, p. 1.

———. February 23, 1998. U. S. forecasts bumper corn and soybean crops and meat glut, p. 3.

———. March 1, 1998. Framers, traders fret over El Niño: If unusual weather cuts harvest, prices of corn and other crops could blossom, p. 1.

———. March 2, 1998. El Niño's fawning ways, p. 3.

———. March 5, 1998. Springlike February warms up sales for major U. S. retailers, p. 3.

———. March 6, 1998. El Niño has another balmy effect: Though some stores—in storm-stricken areas—suffered—many retailers benefited from warm weather last month, p. 1.

———. March 10, 1998. About that mild winter, p. 14.

———. March 20, 1998. In the short run, analysts see strength, p. 1.

———. March 23, 1998. Twice-cursed Grand Forks area gives thanks to El Niño winter, p. 8.

———. March 30, 1998. Low mortgage rates, warm weather lead to record home sales, p. 3.

———. April 1, 1998. Mild winter helps salt away funds, p. 1.

———. April 5, 1998. Sellers market; brisk winter for home sales signals rapid pace for spring, p. 7C.

Cincinnati Inquirer. October 3, 1997. El Niño threatening higher banana prices, p. C09.

———. January 7, 1998. Warm winter foreshadows buggy spring, p. E01.

———. March 4, 1998. Business Notes, p. B03.

———. March 20, 1998. El Niño lion tamed winter, but likely more lamb in spring, p. B1.

———. March 22, 1998. Travel destinations fall victim to El Niño, p. T09.

———. March 26, 1998. El Niño prediction uncertain—lots more known, but not the 'why'? p. A15.

———. April 4, 1998. Job decline first in two years—cooling pace averts inflation, p. B16.

Cleveland Plain Dealer. September 18, 1998. El Niño set to boost grain prices, p. 1C.

———. November 9, 1997. Amid mild forecasts utilities prepare for worst, p. 6B.

———. February 2, 1998. Nothing is average about El Niño effect, p. 1A.

———. February 13, 1998. For fishermen no ice doesn't no dice, p. 13D.

———. March 1, 1998. When winter went AWOL: February set records for least snowfall and warmest temperatures, p. 1A.

———. March 31, 1998. Rare tornadoes smite Minnesota, p. 7A.

Climate Prediction Center. June 10, 1997. *ENSO Advisory*. Washington, DC: National Weather Service, National Oceanic and Atmospheric Administration.

———. August 13, 1997. *ENSO Advisory*. Washington, DC: NWS, NOAA.

———. November 1997. *Special Climate Summary, 97/3*. Washington, DC: NWS, NOAA.

———. February 12, 1998. *Prognostic Discussion of SST Forecasts*. Washington, DC: NWS, NOAA.

Clinton, W. J. March 1, 1998. It Keeps Going and Going. *ABC News*.

Cornell, B. October 29, 1998. Tropicana squeezing up orange juice price. *Chicago Tribune*, p. C1.

Daley, W. M. February 13, 1997. *El Niño still going strong: Impacts felt around the country*. Washington, DC: NOAA.

Davis, J. November 17, 1997. There is no shortage of misconceptions associated with El Niño: Loose talk is influencing consumer behavior and businesses in quaint ways. *Minneapolis Star Tribune*, p. 1D.

Des Moines Register. October 26, 1997. Natural gas stock may very well be the only thing El Niño won't affect, p. 1.

———. November 3, 1997. Do you think Des Moines will have a bad winter because of El Niño? p. 3.

———. November 16, 1997. Will El Niño be a fair or foul weather friend? p. 1.

———. January 14, 1998. Slow going: It's not snowing, p. 1.

———. February 7, 1998. El Niño, winter storms continue to pound California and Kentucky, p. 3.

———. February 19, 1998. Iowans likely to face cooler and wetter spring, p. 1.

———. February 23, 1998. El Niño produces $39 million windfall for MAE customers, p. 2.

———. March 8, 1998a. El Niño system means mayhem for gardeners, p. 05.

———. March 8, 1998b. Deadline for crop insurance approaching, p. 1.

———. March 20, 1998. El Niño's little sister means dry summer, p. 1.

Detroit News. September 22, 1997. Investing markets reacting to any old noise, p. 9.

———. October 30, 1997. El Niño has become the perfect scapegoat for everything, p. E8.

———. December 10, 1997. Michigan ski industry is vibrant, p. F8d.

———. January 8, 1998. Mild weather keeps business hot, p. C3.

———. February 6, 1998. Thank El Niño for those lower heating bills, p. A1.

———. February 15, 1998. Snowmobilers perish in bay, p. A1.

———. February 23, 1998. El Niño kicks up allergy season: wet mild winter threatens havoc with pollen and molds, p. A5.

———. March 3, 1998. Local produce buyers say El Niño is hurting crops, p. F1.

———. March 20, 1998. Farewell to winter that wasn't: Warmth hurt businesses that depend on snow, but it helped consumers and environment, p. E1.

———. March 12, 1998. You could call it El Niño dividend, p.B1.

————. April 2, 1998. Mild weather lets athletes get outdoors and lower times, p. D10.

————. April 15, 1998. Inflation barely detected in first quarter, p. B3.

————. June 5, 1998. Retailers make gains in May, p. 3.

————. June 23, 1998. Fish flies are back, slimier than ever, It's El Niño, p. D1.

Dole, R. M. 1998. Effect of the 1997–98 El Niño on major U. S. weather events. *EOS, Transactions AGU*, 79 (45), F27.

Farm Week. March 6, 1998. Weather uncertainties loom over '98 planting, p. 6.

————. April 6, 1998. Moderate winter, insect adaptation bad signs? p. 4.

————. January 10, 1999. Carryover outlooks deepen winter blues, p. 7.

Federal Emergency Management Administration. August 12, 1997. *Strong El Niño could disrupt winter weather patterns*. Washington, DC: FEMA.

————. September 12, 1997. *U. S. residents urged to prepare in advance for potentially heavy rains and flooding expected to accompany this year's powerful El Niño*. Washington, DC: FEMA.

————. February 22, 1998. *El Niño '98 storms recovery summary*. Washington, DC: FEMA.

————. March 4, 1998. *Preparing for the El Niño '98 storms: A compilation of successful mitigation projects*. Washington, DC: FEMA.

Federal Reserve Board. February 18, 1998. U. S. industrial output turns flat in January: El Niño gets the blame because warmer than normal weather hurt production at gas and electric utilities. *Minneapolis Star Tribune*, p. 4D.

Financial Times. July 28, 1997. Losses from El Niño 1982–1983, p.1.

Friday, E. W. March 1, 1998. It keeps going and going. *ABC News*.

Fonda, R. April 1, 1998. Mild winter helps salt away funds. *Chicago Tribune*, p. 1.

Glantz, M. H. 1998. El Niño forecasts: Hype or Hope? *Network Newsletter*, 13, 1.

Gray, W. November 29, 1997. Hurricane season oddly quiet in 1997. *Champaign-Urbana News Gazette*, p. 5.

Great Lakes Update. April 1998. Detroit, MI: U.S. Army Corps of Engineers.

Guimares, R., Hefner, F., and Woodward, D. 1993. Wealth and income effects of natural disasters: An econometric analysis of hurricane Hugo. *Review of Regional Studies*, 23, 97–114.

Hall, J. M. September 11, 1997. *Testimony before Subcommittee on Energy and Environment*. U.S. House Committee on Science, Washington, DC.

Hunt, A. G. 1999. Understanding a possible correlation between El Niño occurrence frequency and global warming. *Bulletin Amer. Meteoro. Soc.*, 80, 297–300.

Illinois State Water Survey. March 1998. *Illinois Water and Climate*. Summary.

Kasrield, P. May 1, 1998. Low inflation, robust growth, economy turns out "phenomenal" first quarter. *Chicago Tribune*, p. 1

Kerr, R. A. 1999. Big El Niños ride the back of slower climate change. *Science*, 283, 1108–1109.

Kirtman, B. March 1998. *Experimental Long-Lead Forecast Bulletin*. Calverton, MD: Center for Ocean-Land-Atmosphere Studies.

Kocin, P. J. 1997. Some thoughts on the societal and economic impacts of winter storms. *Proceedings Workshop on the Societal and Economic Impacts of Weather*. Boulder, CO: NCAR, 55–60.

Le Comte, D. 1999. A warm, wet, and stormy year. *Weatherwise*, 52, 19–28.

Leetma, A. March 25, 1998. El Niño's winter no more costly than others. *USA Today*, p. 1.

Michaels, P. April 19, 1998. Beware, was El Niño combined with global warming? No! *Los Angeles Times*, p. 5A.

Minneapolis Star Tribune. October 22, 1997. Blame El Niño, p. 8B.

———. October 27, 1997. A storm of superlatives: Taking El Niño to the bank, p. 5A.

———. November 17, 1997. There is no shortage of misconceptions associated with El Niño: Loose talk is influencing consumer behavior and businesses in quaint ways, p. 1D.

———. March 18, 1998. Goodbye El Niño: Although El Niño produced an unseasonably warm winter, that doesn't mean a sizzling summer. Meteorologists foresee a lot of normal weather ahead, p. 14A.

———. March 29, 1998. Get ready for allergy season, p. 5E.

———. May 29, 1998. El Niño bugfest unlikely, infestations of mosquitos not forecast, P. 1B.

National Association of Realtors. 1998. *Home Sales New Records for the Winter*. Washington, DC: NAR.

National Climatic Data Center. July 1998. *California Flooding and Florida Tornadoes, February 1998*. Asheville, NC: NAR.

National Highway Traffic Safety Administration. 1999. *Data on Vehicular Deaths*. Washington, DC: U. S. Department of Transportation.

National Research Council. 1999. *The Costs of Disasters: A Framework for Assessment*. Washington, DC: National Academy of Sciences Press.

National Safety Council. 1999. *Road Safety Statistics*. On Internet at http://www.national-safety-council.ie/road/stat., p. 2.

National Weather Service, 1997–1998: Storm Data. Issues for December 1997 and January–March 1998. Washington, DC: Government Printing Office.

National Oceanic and Atmospheric Administration (NOAA). March 9, 1998. *January and February warmest and wettest on record*. Washington, DC: NOAA.

———. June 30, 1998. *Information on Florida's weather that led to fires*. Washington, DC: NOAA.

Pearce, J., and Smith, J. March 20, 1998. Farewell to winter that wasn't: Warmth hurt businesses that depend on snow, but it helped consumers and environment. *Detroit News*, p. E1.

Peterlin, A. April 15, 1998. Weather problems. *Detroit News*, p. C1.

Pielke, R., and C. W. Landsea. 1998. Normalized hurricane damages in the U. S. 1925–1995. *J. Weather Analysis and Forecasting*, 7, 621–631.

Property Claim Service. October 1998. *Catastrophe Record by Year 1998*. New York: PCS.

Reuters News Service. March 17, 1998. *Effects of El Niño*. (Internet).

Richards, F., Dione, J., and Kluck, D. 1998. *National Hydrologic Outlook: Spring Flood Potential*. Silver Spring, MD: Office of Hydrology, National Weather Service.

Ross, T., Lott, N., McCown, S., and Quinn, D. 1998. *The El Niño Winter of '97–'98*. NCDC Technical Report 98–02. Asheville, NC: NOAA.

Sacramento Bee. October 15, 1997. L. A. summit seeks joint approach to El Niño, p. 1.

St. Louis Post-Dispatch. February 22, 1998. Early spring fishing may end up better this year, p. F11.

———. March 9, 1998. Another El Niño tale: The eagles haven't landed here, p. 6.

———. March 25, 1998. Despite heavy rains, no great flood of '98 is likely, experts say, p. B1.

San Francisco Chronicle. August 14, 1997. Devastating El Niño forecast, p. 1.

Shaefer, M. October 30, 1997. *Testimony before Subcommittee on Water and Power*, U. S. House Committee on Resources, Washington, DC: GPO.

Skinner, B., Lamb, P., Richman, M., and Snow, J. October 17, 1997. *The 1997–1998 El Niño—Possible Impacts on the Property Insurance Industry*. Norman: Cooperative Institute for Mesoscale Meteorological Studies, University of Oklahoma.

Stamper, J. March 7, 1998. Gasoline prices decrease. *Seattle Times*, p. 4.

Stread, R., and Thomason, E. March 6, 1998. Some stocks and investors weather El Niño. *St. Louis Post-Dispatch*, p. C15.

Time. March 23, 1998. El Niño's (achoo) allergies. 151, 34.

U.S. Department of Commerce. March 17, 1998. *Construction of new Homes in February*. Washington, DC: Author.

———. April 1998. Inflation barely detected in first quarter. *Detroit News*, p. B3.

———. May 1998. Low inflation and robust growth. *Chicago Tribune*, May 1, p. 1.

U.S. Department of Labor. March 7, 1998. Mild weather helps to build job growth. *Chicago Tribune*, p. 1, Section 2.

———. March 14, 1998. Wholesale prices not up for fifth month in a row: Asian crisis and El Niño force producers to charge less for goods. Washington, DC: Associated Press.

University of Illinois. November 4, 1997. *El Niño does not increase the risk of large storms on the Great Lakes*. Champaign: Author.

University of Michigan. October 24, 1997. *UM researchers find links between El Niño and weather in the Great Lakes region*. Ann Arbor: University of Michigan.

USA Today. March 2, 1998. El Niño has absolutely crazy effect on business, p. B1.

———. March 10, 1998. Gas prices plunge to historic low, p. 1.

Waetke, K. February 23, 1998. El Niño produces $39 million windfall for MAE customers. Lower energy prices also help reduce natural gas heating bills by 16 percent this winter. *Des Moines Register*, p. 2.

Wall Street Journal. February 6, 1998. Retailers say sales in January rose strongly, p. B4.

———. March 6, 1998. Retailers top forecasts for sales in February, p. B4.

———. April 10, 1998. Retailers post modest March gains, p. B4.

———. December 7, 1998. Slide in energy prices may not be done, p. C1.

West, C. T. and Lenze, D. C. 1994. Modeling the regional impact of natural disasters and recovery: A general framework and application to hurricane Andrew. *International Regional Science Review*, 17, 121–150.

Wisely, R. June 5, 1998. Retailers make gains in May. *Wall Street Journal*, p. B3.

Wilson, W. T. September 22, 1997. Investing: Markets reacting to any old noise: El Niño among other things seen as latest threat to stability. *Detroit News*, p. F9.

Wolter, K. October 28, 1997. El Niño gave blizzard much of its strength. *USA Today*, p. 4.

7

 ## Policy Responses to El Niño 1997–1998

Implications for Forecast Value and the Future of Climate Services

ROGER A. PIELKE JR.

El Niño 97–98 will be remembered as one of the strongest ever recorded (Glantz, 1999). For the first time, climate anomalies associated with the event were anticipated by scientists, and this information was communicated to the public and policy makers to prepare for the "meteorological mayhem that climatologists are predicting will beset the entire globe this winter. The source of coming chaos is El Niño . . ." (Brownlee and Tangley, 1997). Congress and government agencies reacted in varying ways, as illustrated by the headlines presented in Figure 7-1.

The link between El Niño events and seasonal weather and climate anomalies across the globe are called teleconnections (Glantz and Tarlton, 1991). Typically, during an El Niño cycle hurricane frequencies in the Atlantic are depressed, the southeast United States receives more rain than usual (chapter 2), and parts of Australia, Africa, and South America experience drought. Global attention became focused on the El Niño phenomenon following the 1982–1983 event, which, at that time, had the greatest magnitude of any El Niño observed in more than a century. After El Niño 82–83, many seasonal anomalies that had occurred during its two years were attributed, rightly or wrongly, to its influence on the atmosphere. As a consequence of the event, societies around the world experienced both costs and benefits (Glantz et al., 1987).

Another lasting consequence of the 1982–1983 event was an increase in research into the phenomenon. One result of this research in the late 1990s has been the production of forecasts of El Niño (and La Niña) events and the seasonal climate anomalies associated with them. This chapter discusses the use of climate forecasts by policy makers, drawing on experiences from El Niño 97–98, which replaced the 1982–1983 event as the "climate event of the century." The

FIGURE 7-1. Government agencies and officials reacted to the El Niño–based forecasts with mitigative actions in some states and then followed up government relief with assistance based on the losses caused by the damaging weather, as illustrated here. Federal, state, and local agencies all got into the act in those states where bad outcomes occurred.

purpose of this chapter is to draw lessons from the use of El Niño–based climate forecasts during the 1997–1998 event in order to improve the future production, delivery, and use of climate predictions. This chapter focuses on examples of federal, state, and local responses in California, Florida, and Colorado to illustrate the lessons. A full accounting of the use, misuse, costs, and benefits of the recent event awaits full documentation. Nonetheless, it is not too early to begin consideration of the implications of the 1997–1998 event for how we think about value of climate forecasts and for the future of climate services in both the public and private sectors.

El Niño–related forecasts lend themselves to confusion as scientists predict a range of different phenomena (chapter 4). First, there are predictions of the onset, duration, and intensity of the El Niño (or La Niña) phenomenon itself. Such forecasts are typically associated with the state of the sea surface temperature (SST) in the equatorial Pacific, as explained in chapter 2. Because these forecasts focus on the El Niño phenomenon, they do not lend themselves in a straightforward way to predictions of regional climate anomalies or variations. There is some question as to how well the onset and magnitude of El Niño 97–98 was predicted. According to Michael McPhaden, of the National Oceanic and Atmospheric Administration (NOAA), "I don't think anyone in the community realized how strong it would be. Another great surprise was how quickly it ended" (*New York Times*, February 14, 1999). Ants Leetma, director of NOAA's Climate Prediction Center, stated, "If you look at all the forecast models, there was no consensus at the outset that this would be a major event" (Spotts, 1998). And, according to the World Meteorological Organization (WMO), it "developed more quickly and with higher temperature rises than ever recorded" (WMO, 1998).

But, there is no question that once the event was under way, scientists in the United States did an impressive job of predicting a range of seasonal climate anomalies for the coming fall, winter, and spring. These ranged from excessive rainfall and storms in California and Florida to depressed hurricane activity in the Atlantic basin. Chapters 2 and 5 of this volume document many of these skillful forecasts. Because of this apparent success, the climate community has raised expectations that climate predictions will become more and more skillful (e.g., Uppenbrink, 1997). As the scientific community begins to move toward "operational forecasting," it will be important for the nation to develop a robust program of *climate services*. That is, for society to realize the benefits related to the use of climate predictions, attention will have to be paid to both the production of forecasts and the use of forecasts by decision makers. This chapter focuses on how we might begin to evaluate a program of climate forecasts with the goal of improving their use and, ultimately, their benefits to society. Chapter 5 illustrates many of the problems faced by decision makers in deciding whether to use the 1997–1998 forecasts.

FORECAST GOODNESS: USE, MISUSE, AND NONUSE OF PREDICTIONS

Predictive information has become fundamental to a wide range of decisions. Although weather forecasts are the most ubiquitous, in recent years climate forecasting has gained prominence, but only as experiments, not as full-fledged "operational" products. The scientific community has for the most part labeled predictions of seasonal climate anomalies associated with El Niño 97–98 as an example of truly successful operational climate forecasts. Consequently, both scientists and decision makers have asked how good the climate forecasts of

1997–1998 were and what should be expected for the future (see, e.g., Baker, 1998; Barnston et al., 1999).

Decision makers have long faced challenges in the evaluation of predictions that they use, even as demand for predictive information grows (Sarewitz et al., in press). Because decisions are by their nature forward looking, decision makers seek knowledge of the future. Similarly, scientists and other experts have taken advantage of growing computer power and enormous databases to produce a growing range of predictive information. This growth in the predictive enterprise and its value can become problematical if predictions are misunderstood, misleading, or misused.

Society's ability to answer the question "What is a good forecast?" is central to important policy decisions related to weather and climate, especially protection of life and property from extreme events. Being able to answer this question is also important with respect to policy makers' ability to effectively allocate resources between support for forecasting and weather-related research and alternative means of benefiting decision makers (e.g., hazard mitigation policies). On a broader scale, the answer to this question is central to the ability of policy makers to allocate scarce resources between weather-related concerns and the many others demands for federal support.

Murphy (1993) used the phrase "forecast goodness" to refer to the broad societal benefits of a forecasting process. He used the broad term because evaluation of forecasts requires consideration of a range of dimensions. This chapter pays explicit attention to two of these dimensions: forecast skill and forecast value (Roebber and Bosart, 1996). The former refers to the scientific success of a forecast and the latter to the economic benefits associated with the forecast. For a forecasting process to be considered a success, it must show both skill and value.

Prediction as a Process, Not Simply a Product

Before discussing the concepts of skill and value, it is important to present the notion of a *prediction process*. Too often we think of forecasts simply as products and neglect the broader context in which predictions occur. A growing body of experience suggests that the successful use of predictions depends more on a healthy process than on just "good" information (Sarewitz et al., in press; see also chapter 5). The prediction process can be thought of as three parallel processes of decision (cf. Glantz and Tarlton, 1991):

> *Prediction Process*: Includes the fundamental research and observations, as well as forecasters' judgments and the organizational structure that go into the production of operational forecasts.
>
> *Communication Process*: Includes both the sending and the receiving of information. A classic model of communication is *who says what to whom, how, and with what effect.*

Choice Process: Includes the incorporation of weather forecast information in decision making. Of course, decisions are typically contingent on many factors other than forecast information.

Often, some persons mistakenly ascribe a linear relationship to the processes. From the perspective of societal benefits, these three processes are better thought of as components of a broader *prediction process*, with the subprocesses taking place in parallel, with significant feedbacks and interrelations among them.

A central, but often overlooked, point is that success, according to the criteria of any subset of the three processes, does not necessarily result in benefits to society (chapter 5). A technically skillful forecast that is miscommunicated or misused can actually result in costs to society (see the case of forecasts of flood crests in Pielke, 1999). Similarly, effective communication and use of a misleading forecast can lead to decisions with undesirable outcomes. For society to realize the benefits of the resources invested in the science and technology of forecasting, success is necessary in all three elements of the forecast process: prediction, communication, and choice. Furthermore, success requires healthy connections between each of the elements; they cannot be considered in isolation.

Defining Skill and Value.   Because the forecast process comprises multiple elements, there is no single measure that captures the societal "goodness" of a forecast process (Murphy, 1993). Instead, multiple measures are needed to evaluate the technical, communication, and use dimensions of forecasts. Typically, policy makers have focused attention on the simple economics of forecasts in order to determine a "bottom-line" assessment of value. Social scientists have studied the communication process (e.g., warnings), and physical scientists have evaluated forecasts according to technical criteria like skill scores and critical success indexes. These different foci are clearly important; however, the segregation of evaluation tasks has meant that no one is responsible for evaluation of the entire forecast process. The net result is that we tend to view the forecast process as a series of independent tasks, rather than as interrelated decision processes, resulting in partial evaluations at best.

How does one know if a forecast is a good one? The answer is a bit trickier than one might think. Consider the case of early tornado forecasts (Murphy 1996). In the 1880s, a weather forecaster began issuing daily tornado forecasts in which he would predict for the day "tornado" or "no tornado." After a period of issuing forecasts, the forecaster found his forecasts to be 96.6 percent correct—a performance that would merit a solid "A" in any grade school. But others who looked at the forecaster's performance discovered that simply issuing a standing forecast of no tornadoes would result in an accuracy of 98.2 percent! This finding suggested that, in spite of the high degree of correct forecasts, the forecaster was providing predictions with little *skill*—defined as the improvement of a forecast over some naïve standard—and in fact his forecasts could result in costs rather than benefits.

Simply comparing a prediction with actual events does not provide enough information with which to evaluate the prediction's performance. A more sophisticated approach is needed.

Scientists have a range of techniques that they use to assess the skill of a forecast (Murphy, 1997). One way to evaluate skill is to compare the prediction with some baseline. Climatology, that is, historical weather information aggregated over time and space, provides such a baseline because it provides the best estimate of the future occurrence of weather events, absent any other information. Thus, a forecast is considered skillful if it improves on a prediction based on climatology. For instance, assume that the average high temperature over the past 100 years in London on September 6 is 10°C (i.e., the climatological mean for that date). Absent any other information, the best prediction of the temperature in London for the next September 6 is thus 10°. Any forecast for that particular day would be considered skillful if it were to improve on climatology in comparison to the actual temperature recorded on that date. For forecasts that are probabilistic, rather than categorical, the evaluation of skill can be somewhat more complicated, but it adheres to the same principles (see Murphy, 1997).

Another way to evaluate a forecast is to assess whether it had *value* to a decision maker, where value is typically measured in economic units. Murphy (1997) notes that a forecast has no value if it does not affect a decision. Consequently, even a highly skillful forecast might not have value. For instance, it might be the case that the decision maker has no flexibility to use the forecast or that the prediction arrives in a form or at a time irrelevant to the decision context. For example, a prediction of seasonal hurricane activity is of little value to the reinsurance industry if the forecast is made after insurers have already negotiated contracts with their clients (Malmquist and Michaels, 1999). It is also possible that a forecast could have negative value; that is, it could result in costs rather than benefits. This could occur if the forecast is misunderstood, miscommunicated, or misused in some fashion. In the flooding of the Red River of the North, in 1997, many decision makers failed to appreciate the uncertainty present in what turned out to be a skillful forecast of the flood crest (Pielke, 1999). As a result, their decisions based on that forecast arguably led to costs, rather than benefits.

In short, a skillful forecast is not a sufficient condition for a forecast to have value (see Stewart, 1997, on theoretical value considerations, and Sarewitz et al., in press, for a set of practical case studies). In some cases, a skillful forecast is not even *necessary* for the forecast to have value (on this issue, the history of earthquake prediction is instructive; see Sarewitz et al., in press). The relationship of forecast skill and its value to society is complex and context-sensitive.

Figure 7-2 is a 2 × 2 matrix showing forecast skill along one axis and forecast value along the other axis. The resulting four quadrants show four possible outcomes: (1) skill and value, (2) no skill and value, (3) skill and no value, and (4) no skill and no value. To understand the aggregate value of a particular climate

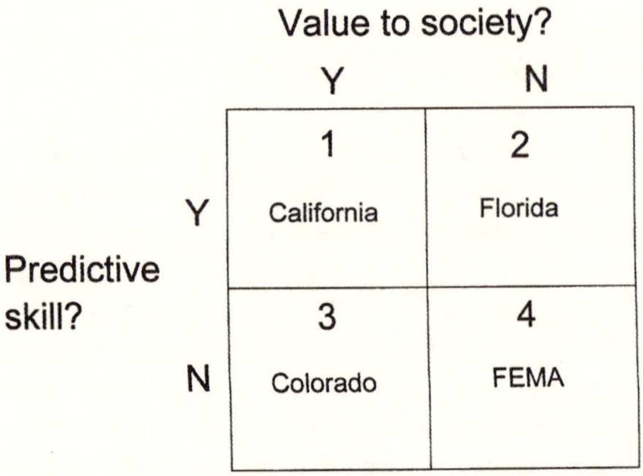

Value to society?

	Y	N
Y	**1** California	**2** Florida
N	**3** Colorado	**4** FEMA

Predictive skill? (row label)

FIGURE 7-2. The relationship between forecast skill and forecast value, with examples of each outcome, related to cases of predictions described in the text.

forecast, one must carefully consider the full range of skill and value outcomes. For a mature forecasting process, a policy goal might include seeking certain levels of success on a systematic basis with respect to both skill and value. But, for a new forecasting process, it is likely that outcomes will occur in each quadrant, providing a significant body of experience on which to improve performance of the system. In other words, the purpose of such a typology is to identify outcomes that fall into categories 2, 3, and 4 in order to modify the forecasting process in such a way as to shape future outcomes toward one of skill and value (category 1). The case studies of predictions described in the next section are identified in each of the four boxes on Figure 7-2.

EXAMPLES OF FORECAST APPLICATIONS DURING THE 1997–1998 EVENT

It is important to recognize that a comprehensive evaluation of the skill and value of any forecasting process requires a considerable research effort. Consider that for the United States, the National Oceanic and Atmospheric Administration (NOAA) issued seasonal predictions for the entire country that were broadly disseminated through the media to millions of potential decision makers. The purpose of presenting the four sets of examples is not to provide a bottom-line assessment of the "benefits" of El Niño–related predictions. It is, instead, to illustrate a framework for feedback to the scientific and policy communities about how climate services can be improved in the future and also to suggest a mechanism for their evaluation.

Skill and Value: Responses in California

The case of predictions of seasonal anomalies in California is perhaps the most widely publicized (chapters 3 and 6). A growing body of evidence suggests that these forecasts were effectively used in many instances by decision makers to prepare for exceptionally heavy precipitation (chapter 5). For instance, James Baker, administrator of NOAA, testified before Congress in October 1998 about the responses his agency sought to stimulate on the basis of the prior year's El Niño forecasts and observations of the event. "There is strong evidence that the $500+ million in damages in California in the first three months of 1998 compared to $1.8 billion in the first three months of '95 and '97 was due to better preparation by state and local officials" (Baker, 1998).

Decision makers indeed conducted a wide range of preparatory actions in California to El Niño–related climate predictions. For instance:

1. On October 6, 1997, Governor Pete Wilson convened an "El Niño Summit" of fourteen agency secretaries and department directors to go over state preparations (California Resources Agency, October 6, 1997). The governor signed legislation that earmarked $7.5 million specifically for El Niño preparations, including funding for levee work and increased staff to monitor flood forecasts and reservoir operations. The governor also signed an executive order to coordinate the state response, set in motion planning for seven regional briefings for the months of October and November 1997, and submitted a letter to President Clinton requesting federal aid for various projects. The summit received tremendous media attention, with more than 100 media sources in attendance (Federal Emergency Management Agency, 1997). As a result of this high-level attention, Douglas Wheeler, Secretary of Resources for the State of California, stated that "El Nino preparation is now a cabinet-level priority" (Showstack, 1997).
2. On October 14, 1997, a second, highly visible "El Niño Summit" was organized by the White House, at the behest of California Senator Barbara Boxer (C. Visserig, personal communication). This summit garnered tremendous national attention and led to a cascading series of preparatory actions by the State of California and its communities. For instance, on December 8, 1997, the Association of Bay Area Governments, the Governor's Office of Emergency Services, and the U.S. Geological Society held a conference for local officials to help them prepare for expected heavy precipitation.
3. The Federal Emergency Management Agency (FEMA) took a number of actions in California. These included streamlining the processing time for "typical small disaster survey reports"; prioritizing, with

the Governor's Office of Emergency Services, flood-related projects; developing an expedited review process for outstanding hazard-related assistance and projects to clear the way for future needs; and initiating various public affairs operations (Armstrong, 1997). Other federal agencies taking actions in California included the Environmental Protection Agency which gave attention to toxic cleanup sites and landfills, and the Army Corps of Engineers, which worked on flood control projects (Barnum, 1997).

4. As a result of the various summits, briefings, and constant media attention, many decision makers in California took action in preparation for anomalous heavy rainfall. According to NOAA's California Pilot Project on the Use of Climate Forecast Information, the Orange County Sanitation District (OCSD) began taking action in the summer of 1997 (NOAA, March 10, 1998; see also Epstein, 1997). The district manages wastewater collection, treatment, and disposal service for twenty-three of the county's thirty-one cities. Actions taken by the district included plugging holes in manhole covers and public education about water usage during periods of high precipitation. Both of these activities can contribute to lessening the total volume of water flowing through the wastewater systems, thereby reducing the chances that system capacity will be exceeded. The district also developed a high-flow emergency plan that would be triggered by forecasts of pending heavy precipitation. According to the OCSD, the advance planning paid off during heavy rains related to the El Niño event. "Approximately $300,000 worth of damage to OCSD's facilities resulted from the high flows of the winter storms of 1995. Flows of similar magnitude during December 1997—February 1998 were managed more effectively, due in part to the advance notice that the OCSD received of major precipitation events, and resulted in no major facility damage" (NOAA, March 10, 1998).

5. Other examples of the multitude of preparatory actions in California in the fall of 1997 included these actions: San Diego lifeguards received special training for river rescues; the city staff of Malibu was trained in special disaster-response drills; and Southern California Edison trained 500 employees to use a "multimillion-dollar power failure management system" (Scheeres, 1997). Chapter 5 and other sources (Barnum, 1997; CRA, October 20, 1997; Epstein, 1997; *USA Today*, 1997) provide additional examples of the wide range of preparations that occurred in California.

The use of climate predictions in California seems very much like a best-case scenario, although flooding was still a major problem, as illustrated in Figure 7-3. But, important lessons for wise use of predictions can still be learned.

FIGURE 7-3.   The wet winter in California brought extensive flooding, as illustrated by this February photograph. (Courtesy Robert A. Eplett, California Office of Emergency Services)

First, the effective use of a skillful prediction does not always mean the creation of benefits. In one instance, prediction-based preparations caused some controversy. In Los Angeles, efforts to clear the trees and brush from several river channels ran into opposition from environmental groups seeking to preserve habitats for riverine species (*New York Times*, November 9, 1997). Predictive information led to costs, rather than benefits, from the perspective of those wanting to preserve the flora and fauna that had come to occupy the Los Angeles flood ways. The lesson here is that in addition to aggregate outcomes, often the value of predictive information also includes considerations of who benefits and who does not. More precisely, for each case, in the regional applications of forecasts, there will be a set of outcomes that could be described by the four quadrants of Figure 7-2. Hence, placing the responses in California in just one quadrant oversimplifies the richness of the responses.

Second, forecasters must be careful to manage the expectations of decision makers. The case of California is a success story, and forecasters have not been shy in touting their success. But what would have happened if the storms had failed to materialize? Michael Glantz, of the National Center for Atmospheric Research (NCAR), commented, in September 1997, that "if southern California doesn't get slammed, there's going to be hell to pay" (Frankowski, 1997). The connection of El Niño with seasonal climate anomalies in California is well documented. But, all predictions of climate are necessarily probabilistic, meaning that there will be times when a skillful prediction is made, yet the event being predicted does not materialize. In a case like California, an important challenge for forecasters might well be to manage expectations following realized forecasts.

No Skill and Value: The Colorado Blizzard

As it became apparent that El Niño 97–98 was going to be the most intense such event in fifteen years, residents of Colorado began to compare its potential impacts with those of the 1982–1983 event. Along Colorado's Front Range, where Denver is located and three million persons reside, this meant references to the largest blizzard in recent memory. "This year's conditions appear to parallel those in 1982–83 . . . when Denver experienced the Blizzard of '82" (Schrader, July 26, 1997). Figure 3-4, from a Denver newspaper in September, helps explain the local situation. But, some scientists expressed caution about drawing too close an analogy to only one previous El Niño. Several local atmospheric scientists were quoted in July 1997 as warning that the developing El Niño could bring "drought or deluge" to Colorado (Schrader, July 26, 1997). According to Michael Glantz, a scientist at the nearby NCAR, the comparisons with the 1982 blizzard is "starting to become a myth. There's an aura starting to develop around the 82–83 event" (Frankowski, 1997).

As a result of the various ominous climate forecasts that appeared in the media, Denver officials met with scientists at the NOAA offices in Boulder in mid-October 1997 for a briefing on what winter conditions might be expected (Flynn, 1997). The Denver Mayor's Office was sensitive about any hint of the potential for a blizzard, as Mayor Bill McNichols, who was in office in 1982, was widely blamed for the city's inability to handle the tremendous 1982 snowfall and was subsequently voted out of office (Martinez, 1997). The message of the NOAA scientists to Denver city officials was summarized as "Denver could experience a drier-than-normal winter followed by a wet spring, when the region could be buried under heavier-than-normal precipitation" (Martinez, 1997). In response, Denver Mayor Wellington Webb announced, on October 23, that he had rearranged the city's snow-plowing plans, made plans to stockpile salt and de-icing chemicals, and budgeted additional funds for overtime snowplow operations (Flynn, 1997; Martinez, 1997). The new plan concentrated snowplowing activities on the city's major streets, leaving other streets to be plowed after any major storm subsided.

Just a week following the mayor's announced plans, a major blizzard struck the Colorado Front Range, extending east across the High Plains. More than two feet of snow fell in Denver. The storm was Denver's worst in October since 1923 (Lowe and McPhee, 1997). While the city of Denver seemed well prepared for the storm, the main highway to Denver International Airport became impassible, meaning that passengers were stranded at the airport, which had remained open for much of the storm (Martinez and Hughes, 1997). The mayor and the management of the Denver International Airport faced some brief criticism following the blizzard (Martinez and Hughes, 1997; Snel, 1997). Figure 7-4 shows an example of the local reactions to the blizzard problems beyond the city's boundaries. Within the city of Denver, major throughways were kept open, and the new snowplowing plan was evaluated as a success (Lowe and McPhee, 1997).

After the blizzard, atmospheric scientists offered conflicting opinions as to whether the storm could be attributed to El Niño. After the blizzard, a headline in one Denver paper, the *Rocky Mountain News*, proclaimed "El Niño Not to Blame for Blizzard," while the front page of Denver's other major paper, the *Denver Post*, stated, "El Niño's First Strike" (Lowe, 1997; Weber, 1997). More conflicting statements are found in news stories by Schrader (October 28, 1997) and Lowe (October 30, 1997). Regardless of the scientific controversies, this major early blizzard, which produced major damages in six states, became a key factor in developing widespread belief in and national media acceptance of the reality of the El Niño climate predictions and the event's capability to produce severe weather, a condition explored in chapter 3.

Scientists continued into 1999 to debate whether the storm was directly caused by El Niño. But, from the perspective of policy, the lessons of the Colorado blizzard do not depend on resolving this issue. For purposes of argument,

FIGURE 7-4. The early NOAA announcements about the development of El Niño 97–98 and the FEMA warnings about the coming bad weather led Denver officials to prepare for heavy snowfall. However, not all the necessary preparations were made, and an early October blizzard created a series of problems in the Denver area, as illustrated here. (Reprinted with permission of Ed Stein, courtesy of the *Rocky Mountain News*)

simply assume that the blizzard was unrelated to the El Niño; in other words, assume that the predictions based on the 1982 Colorado blizzard had no skill. Even under these circumstances, a compelling case can be made that the predictions still had value. The reason for this is that the 1997 predictions (and the memory of past problems) motivated decision makers to take action on policy processes that needed fixing—in this instance, the snowplowing routes in Denver. The outcome suggests that city officials were only partly successful in their efforts, as the roads to the airport became impassable.

In short, there are two lessons. First, because there was room for improvement in the city's snowplowing plans, the predictions stimulated a "no regrets" response that would have made sense even without the predictions but probably would not have occurred without the El Niño predictions. Second, even a skillful prediction would not have addressed the city's lack of attention to the roads to the airport. The bottom line is that the strengths and weaknesses of existing decision processes are a key factor in whether or not a forecast leads to societal benefits.

Skill and No Value: Florida Tornadoes

Florida held an El Niño Summit on December 15, 1997, two months after the California summit. But, unlike California's summit, the one in Florida was motivated by actual events, not just predictions. "Since last September, severe weather—believed to be largely the result of El Niño—has caused more than $21 million of property damage in Florida, with 1,100 buildings or homes being damaged or destroyed" (Kleindienst, 1997). The summit was encouraged by FEMA and organized by the Florida Department of Community Affairs in order to increase public awareness about the possibility of freezes, increased rainfall, and "more violent storms, especially over central Florida, with a greater chance of more tornadoes" (Kleindienst, 1997). The summit also had the objectives of (1) securing funding for placing NOAA weather radios in every school by June 1, 1998; (2) promoting 100 percent participation in the National Flood Insurance Program; (3) enhancing interagency communications to protect Florida agriculture; and (4) improving the consistency of severe weather warnings for preparedness efforts.

As had been predicted, Florida did experience unusually severe weather during the winter of 1997–1998, with damages totaling more than $500 million (National Weather Service, 1998). The most prominent event in Florida was the tornado outbreak of February 22–23, 1998, which led to $100 million in damages, the loss of forty-two lives, and more than 260 injuries (NWS, 1998). It was the single greatest loss of life to a tornado event in Florida history.

The potential for severe weather, including tornado outbreaks, was anticipated as early as December 15 at the Florida El Niño summit, on the basis of information provided by the National Weather Service (NWS) office in Melbourne,

Florida. As a result, forecasters underwent special training and provided to the public special studies of severe weather impacts and preparedness. Ironically, the week of February 22–28 had been designated as 1997's annual Florida Hazardous Weather Awareness Week. In addition to the longer-term climate outlooks, the specific event was well predicted several days in advance by the NWS's Storm Prediction Center in Norman, Oklahoma. Furthermore, this information was received by the Florida Department of Emergency Management, the Orange County emergency manager, and the Winter Garden city manager. As the storms unfolded, tornado warnings (which are issued for specific tornadoes) were issued with accuracy at levels much higher than national averages. In particular, the four counties that experienced the fatalities were provided average lead times of twenty-three minutes before the tornadoes struck.

The apparently skillful predictions coupled with the tremendous losses led one official in the Florida Department of Emergency Management to wonder "if the forecasts had any value at all?" (C. Fugate, personal communication). There are several possible answers to such a query. One is that the predictions were in fact effectively used by a range of decision makers but that the loss statistics reflect casualties and damages, rather than lives saved and damages avoided. While it is certainly possible that casualties were reduced and damages mitigated because of the predictions, a more satisfactory explanation can be gleaned from the findings of the NWS Service Assessment following the event. The Service Assessment (NWS, 1998) identified "problem areas" in the forecast process.

The overarching problem in this case was that, "even though advance watch and warning information was available and a severe weather effort has existed . . . numerous residents failed to receive or respond properly to the warnings." This occurred in part because the storms struck late at night but also because of technical problems with the dissemination and receipt of warnings. The NWS assessment team also expressed concern that the extended period of heightened awareness served to "desensitize" people to the threatening conditions. Also, the team found that Floridians focused to a much greater extent on hurricane preparedness rather than on tornado preparedness. Most of the deaths (40 of 42) occurred in mobile homes or recreational vehicles, suggesting that structural performance was a factor in the distribution of casualties. What these findings also suggest is that, even with the provision of skillful predictions, the broad forecast process (including long-term considerations such as structural integrity and the provision of safe refuges) did not perform as well as it might have.

It is impossible to answer with certainty how many (or even if) lives would have been saved (or how much in losses avoided) if the broad forecast process had worked more effectively. It is clear in this case that even a highly skilled prediction does not necessarily lead to highly valued outcomes. It is even possible that a skillful prediction might lead to costs rather than benefits, as was the case in the use and misuse of flood crest predictions during a major 1997 flood (Pielke, 1999). The case of the tornado outbreak in Florida in February 1997 should serve as an important lesson—skillful predictions, by themselves, are an insufficient condition for society to realize benefits. An effective forecast process is necessary for the predictive information to be well used and for benefits to result.

No Skill and No Value: El Niño Hype

As the rapidly warming sea surface temperatures of the Pacific in May 1997 confirmed the onset of a record El Niño event, the public was warned to prepare for "history's most costly weather tantrum" (Spotts, 1998). Others saw El Niño as a scapegoat: "magazines and computer Web sites have bulged with fingers pointing the blame at El Niño when anything went wrong" (Pool, 1998). The salience of the phenomenon among the American public led to El Niño jokes making the rounds of late-night talk shows and appearing in advertisements for firewood, snowplows, and even Italian designer clothing (chapter 3). As Figure 7-5 illustrates, coverage of the El Niño event equaled that for other major national issues in the news. On the one hand, greater awareness of factors that can affect climate variability is a good thing, but, on the other hand, mistaken impressions of what El Niño actually is can lead to misuse of climate forecasts.

El Niño 97–98 clearly led to heightened awareness. Figure 7-6 illustrates a measure of media attention to El Nino in eight Midwestern newspapers. It re-

FIGURE 7-5.   The great media hype and the resulting public attention to El Niño led media officials to feature it alongside other major national news stories. (Reprinted with permission of Ed Stein, courtesy of the Rocky Mountain News)

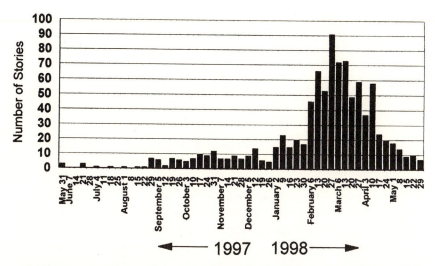

FIGURE 7-6. The graph presents the time distribution of news stories dealing with El Niño published in eight major newspapers.

veals that attention in this region was at its highest during the time of the greatest U.S. impacts, not during the time that the climate forecasts were being issued. A similar pattern was documented by Hare (1998), who found, in a more comprehensive national analysis of media coverage, that

> early stories focused on predictions that the 1997–1998 event would likely challenge the 1982–1983 event as the strongest in modern history. By October [1997] stories started appearing on preparation being undertaken to minimize the impact of El Niño–fueled storms along with estimates of the potential worldwide damage in dollars. In November and December, stories about actual impacts—speculatively linked to El Niño—dominated the stories. . . . In February . . . the number of stories skyrocketed as the southwest and southeast United States both experienced record amounts of rainfall with widespread flooding and damage. In March, reports began to appear on the financial damage resulting from El Niño storms along with stories about the accuracy of El Niño forecasts several months before the onset of the event. . . .

Then, in mid-1998, scientists began to see signs of a coming La Niña event, which began a new cycle of media attention.

The heightened awareness of El Niño that developed in the summer of 1997 did raise questions of a possible overreaction. For instance, consider the following exchange in a congressional hearing on El Niño between Representative Calvert (R-CA) and Tim Barnett, a leading El Niño expert from the Scripps Institution of Oceanography (U.S. House Science Committee, 1997):

CALVERT: Is there any concern of overreaction? You said prepare for the worst. But sometimes we know they can be very dramatic about certain stories. *Time* magazine had a story recently warning of landslides, flash floods, droughts, crop failures during this coming year. Is this a problem? Do you think this is appropriate?

BARNETT: Worries the hell out of me, very frankly. I think what has been lost in the translation from the scientists and the people that do forecasting like myself to the press is the statement of uncertainty associated with it. Although it is given to the press, it oftentimes does not appear and people take it as a certainty that California will be washed away this wintertime.

The many warnings of weather-related dangers issued by FEMA during August-September 1997 as a result of NOAA's El Niño climate forecasts is an example of how the connection between scientists and decision makers broke down because of "El Niño hype." In the spring of 1998, FEMA stated that the El Niño winter of 1997–1998 had not turned out as the agency had expected. From November 1, 1997, to March 31, 1998, the agency had committed about $290 million in response to presidentially declared disasters. This amount is about the same as that committed by FEMA in each of the two previous winters. This statistic surprised FEMA, with one agency spokesperson commenting that "everybody was screaming that El Niño was going to be Armageddon, but our data reveals that's not what it's turned out to be" (Bunting, 1998). The cartoon in Figure 7-7 illustrates in humorous fashion how the public jokingly blamed everything bad on El Niño.

FIGURE 7-7.   This early October 1997 cartoon illustrates how El Niño had invaded the popular culture and become a humorous excuse for all types of personal problems. (Copyright Tribune Media Services, Inc. All rights reserved; reprinted with permission)

Experience shows that El Niño impacts in the United States overall do not mean Armageddon. What El Niño means is that different regions of the country are more or less likely to suffer particular weather events than in La Niña or neutral years. In fact, simply because hurricane damages are much larger during La Niña events, it is probably the case that the nation as a whole experiences fewer overall economic impacts in El Niño years than in La Niña years, as shown in chapter 6. Even so, for those individuals in places likely to experience damaging weather during El Niño events (such as certain parts of coastal California), El Niño events might seem like Armageddon. In aggregate, however, El Niño is better thought of as a shift in the sort of weather impacts that normally occur, rather than as an event with generalized negative impacts for everyone.

As has been shown in the case examples, skillful predictions of El Niño–related impacts can reduce disaster costs. The very fact that El Niño shifts the type and places of impacts that the nation experiences from those seen during non–El Niño years can provide usable information for those who are potentially affected (e.g., Pielke and Landsea, 1999). The summer and fall of 1997 brought a number of policy responses to scientific and media pronouncements that the Pacific Ocean was warming at an unprecedented rate. For instance, the El Niño summit in California focused attention on the possibility of strong coastal storms, and a wide range of decision makers used that information to take preparatory actions. The state of Florida organized its El Niño summit to focus attention on the possibility of extreme weather associated with thunderstorms, but with less apparent success in reducing impacts. In both instances, the advance preparation proved prescient, as both states experienced extreme weather. These examples (and there are many others) provide a better picture of what El Niño information means to the nation than is provided by looking at aggregate impacts. Climate conditions and extreme events like droughts and floods vary regionally across the United States, and consequently losses and gains in any given season, year, or decade vary regionally (Kunkel et al., 1999).

Disaster costs alone do not provide a measure of the value of advance preparation. One might be tempted to conclude that advance preparation efforts had little value because 1997–1998's disaster losses were similar to those of the previous two winters, according to FEMA (Bunting, 1998). This would be a mistake. Because weather impacts are highly random and FEMA's tabulation represents only a subset of the documented impacts (see chapter 6), one cannot compare different years and expect to see indication of the value of preparation. Consider that some in the insurance industry have speculated that better preparation for Hurricane Andrew might have saved $5 billion. But, Andrew would have still been the costliest storm ever (in inflation-adjusted dollars only). Assessing the value of preparation, including advance warnings and forecasts, requires careful attention to the details of specific cases. This demands a considerable investment of time and attention. But, if advance preparation is ever going to become a larger element of the nation's response to extreme weather, then we must know its costs and benefits.

The FEMA announcement, in 1998, that El Niño 97–98 and resulted in fewer losses than expected suggests that perhaps the scientific community was at once too successful and not successful enough in publicizing the coming El Niño. It was too successful to the extent that people came to associate El Niño with Armageddon, and not successful enough to the extent that the public failed to appreciate the subtleties of extreme weather related to El Niño. If the media and the public fail to accurately understand El Niño, much less El Niño predictions, then chances are decreased that even skillful forecasts will have value to decision makers. This is an important lesson to learn as the nation develops skill in seasonal climate forecasting.

THE FUTURE OF CLIMATE SERVICES

Weather Forecasts and Climate Forecasts

The National Weather Service issues on the order of 10 million official weather forecasts annually, to an audience of perhaps 500 million individuals, companies, governments, and other organizations (W. Hooke, personal communication). The number of forecasts and the decision makers who use them provide a body of experience that allows for a stabilization of expectations on the part of decision makers about the goodness of weather forecasts. As a result, you and I know when to take an umbrella when we leave the house, and many sophisticated decision makers have a systematic understanding of the skill, quality, and value of weather forecasts. This also means that a decision made with today's forecast is likely to be substantially similar to a decision made tomorrow with the same forecast, and the next day, and so on.

Such an experiential basis does not exist in the case of climate forecasts. Decision makers do not have an experiential basis for understanding the goodness of forecasts, much less the skill, quality, and value of such forecasts. Thus, decision makers are far more prone to the misuse of climate forecasts. Learning through trial and error will take much longer and will not be as substantial as in the case of weather forecasts, because climate forecasts will be made much less frequently. Decision makers must take care to avoid overinterpretation of the significance of any one forecast. For instance, a realized forecast does not mean that all forecasts will have such success, and an unrealized forecast does not portend future failures. As the nation moves toward the regular provision of more precise climate predictions, the development of a body of shared experience will be critical to the effective use (and to limiting the misuse) of forecasts.

Another, more technical issue is associated with how scientists assess the value of climate forecasts. One common method is to create models of decision routines and use the model to assess the value of improved information. But, because decision makers' understandings of climate forecasts have yet to be stabilized, the decision environment is dynamic and difficult to model. Until such expectations do become stabilized, idealized assessments of climate forecast value are unlikely to reflect accurately the actual value of forecasts. This means that case studies, as shown in chapter 5, are more likely to lead to accurate assessments of the use and value of climate forecasts (on descriptive approaches, see Stewart, 1997).

Uncertainty

Nobel laureate Kenneth Arrow relates a story from his experiences as a weather forecaster in the Air Force during World War II. A group of the forecasters had been assigned the task of forecasting the weather one month in advance. Arrow found that the one-month forecasts developed were no more accurate than forecasts made by simply rolling dice. The other forecasters agreed and asked their superiors to be relieved of this duty. The reply sent back to the forecasters was that "the commanding general is well aware that the forecasts are no good. However, he needs them for planning purposes" (as related by Bernstein, 1997).

The anecdote illustrates the well-known tendency of organizations to gather information, even if the information does not contain useful knowledge (e.g., Feldman and March, 1981). While forecasters cannot control how decision makers use information, they do have control of the information that is provided. Because climate forecasts are necessarily probabilistic, forecasters can communicate in the language of uncertainty and probabilities to give decision makers a rich sense of the seasonal anomalies that they might expect (see Pielke, 1999, for an argument on why probabilistic climate forecasts are important).

The following examples, taken from NOAA press releases prior to El Niño 97–98 and from one issued prior to the 1998 La Niña, suggest that climate forecasters are inconsistent at best in how they communicate uncertainties (compare Barnston et al., 1999).

> *Example 1* Strong El Nino conditions are currently developing in the tropical Pacific. The warm event will bring wetter, cooler weather for the southern half of the United States from November through March, while the northern part of the country from Washington east to the western Great Lakes will experience warmer than normal temperatures. (NOAA, June 17, 1997)

This statement is expressed in categorical, rather than probabilistic terms. No sense of uncertainty is provided.

> *Example 2* California, Texas, and Florida and other states throughout the south are likely to see significant precipitation in the next several months. Based on historical data during past El Niño events, some areas may see as much as 150 to 200 percent of normal. While we tend to view the increased precipitation as a threat, in some instances there are benefits such

as decreasing the chance for wintertime drought in the Southwest and southern Plains and reducing the wildfire danger in Florida. (NOAA, November 5, 1997)

By way of contrast, this statement uses probabilistic language, such as "likely," but does so in a qualitative manner. It also refers to climatology as an indication of the threshold for a skillful forecast.

Example 3 What we're looking at with La Niña is a tilt of the odds toward a colder winter in the Northeast with more snow and perhaps bigger storms than usual. The further north you go, the greater the chance of milder than normal weather with less precipitation. (NOAA, September 9, 1998)

This statement goes even further in its presentation of probabilistic information, but again it does so in a qualitative manner. It does provide a sense of the geographical relationship to probabilities.

Example 4 The long-term climate outlook issued by NOAA today calls for warm and dry conditions in the Southwest during July, August, and September, and these are forecast to continue through the fall and winter . . . the outlooks are also for cool conditions in the northern Great Plains, and dry conditions in interior Washington and Oregon during the next three months. (NOAA, June 22, 1998)

This last example returns to a more deterministic expression of the forecast information. It is not surprising that this wide range of styles in presentation of information to the public resulted in some confusion among decision makers as to what climate forecasts actually mean (see chapter 5).

Before a congressional hearing, Ants Leetma, head of NOAA's Climate Prediction Center, stated that "if the forecasts don't come true, you have me to blame" (House Science Committee, 1997). In this instance, the forecasts, for the most part, did "come true." But they will not always come true. Even so, such a statement is misleading. Just because a single seasonal forecast is unrealized does not mean that a forecast is not "true," just as getting five heads in a row when flipping a coin does not necessarily imply that the coin is loaded.

These various examples suggest that forecasters have an opportunity to improve the use and value of their products by paying more attention to the provision and communication of information in a manner that more accurately reflects the nature of the phenomena being predicted.

CONCLUSIONS

This review of policy responses to the seasonal climate forecasts based on El Niño 97–98 largely supports the notion that the forecast process led to positive outcomes in the use of the forecasts, particularly in the case of California. At the same time, it is clear that there are a great many opportunities to improve the forecast process as the nation moves toward the development of full-fledged climate services. The most important points are these:

- A skillful forecast does not directly lead to value. Each case shows that the decision environment plays a critical factor in the value that is achieved through the use of climate forecasts. The strengths and weaknesses of existing decision processes are a key factor in whether a forecast leads to societal benefits.
- Forecasters must be careful to manage the expectations of decision makers. It is easy to oversell climate forecasts.
- The Colorado snowstorm case illustrates how a prediction can stimulate a "no regrets" response that would have made sense even without the prediction but probably would not have occurred without the prediction acting as a catalyst.
- If the media and the public fail to understand El Niño, much less El Niño predictions, then chances are decreased that even skillful forecasts will have value to decision makers. This places responsibility on forecasters to effectively communicate the significance of climate forecasts to decision makers.
- There is an opportunity for forecasters to improve how probabilistic information is communicated to decision makers. Without an understanding of the probabilistic nature of climate forecasts, decision makers will not make the most effective use of climate forecasts.

The use of climate forecasts will never mirror the use of weather forecasts. But, with attention to the effectiveness of climate services and to the interrelationship between forecast skill and value, the nation has great potential to develop information products that aid decision makers. El Niño 97–98 is but a first step in that direction.

REFERENCES

Armstrong, J. September 11, 1997. *Statement before the Subcommittee on Energy and Environment*. U.S. House Science Committee, Hearings on the State of Science Used in Weather Predictions, House of Representatives, Washington, DC.

Baker, D. J. October 2, 1998. *Testimony before the House Committee on Science*, Subcommittee on Energy and Environment, House of Representatives, Washington, DC.

Barnum, A. October 7, 1997. State scrambles to prepare for El Nino fury: Work on levees, coast, rivers well under way. *San Francisco Chronicle*, p. 1.

Barnston, A. G., Glantz, M. H., and He, Y. 1999. Predictive skill of statistical and dynamical models in SST forecasts during the 97–98 El Niño episode and the 1998 La Niña onset. *Bulletin Amer. Meteoro. Soc.*, 80, 217–243.

Bernstein, R. 1997. *Against the Gods*. New York: Wiley.

Brownlee, S., and Tangley, L. October 6, 1997. The wrath of El Niño. *Newsweek*, p. 17.

Bunting, V. April 4, 1998. Winter damage from El Niño bad but not the worst. *Champaign-Urbana News Gazette*, p. 11.

California Resources Agency. October 6, 1997. Wilson Convenes El Niño Summit to Discuss State's Preparedness for Pending Winter Storms. Sacramento: CRA.

———. October 20, 1997. El Niño Regional Briefing in Long Beach on Potential Weather Impacts, State and Local Preparedness. Sacramento: CRA.

Epstein, E. September 5, 1997. Agencies in Bay area preparing for El Niño: They're stocking up and educating the public. *San Francisco Chronicle*, p. 1.

Feldman, M., and March, J. 1981. Information in organizations as symbol and sign. *Administrative Science Quarterly*, 26, 171–186.

Federal Emergency Management Agency. October 13, 1997. *The making of the El Niño Summit.* FEMA News Desk, on Internet at www.fema.gov.

Flynn, K. October 24, 1997. El Niño's threat of snow alters Denver plow plans. *Rocky Mountain News*, p. 3.

Frankowski, E. September 27, 1997. Weather or not? El Niño: Is this a myth in the making? *Longmont Daily Times-Call*, p. A1.

Glantz, M. H. 1999. *Currents of Change*, 2d ed. New York: Cambridge University Press..

Glantz, M. H., Katz, R., and Krenz, M. (eds.). 1987. *The Societal Impacts Associated with the 1982–1983 Worldwide Climate Anomalies.* Boulder, CO: National Center for Atmospheric Research.

Glantz, M. H., and Tarleton, L. F. 1991: Mesoscale Research Initiative: Societal Aspects. *Report of the Workshop Held 10–11 December 1990.* Boulder, CO: Environmental and Societal Impacts Group, NCAR.

Hare, S. R. October 6, 1998. Recent El Niño brought downpour of media coverage. *EOS:* 34, 481.

Kleindienst, L. December 16, 1997. El Niño making its presence known throughout the state. *Fort Lauderdale Sun-Sentinel*, p. 1A.

Kunkel, K. E., Pielke, R. A. Jr., and Changnon, S. A. 1999. Temporal fluctuations in weather and climate extremes that cause economic and health impacts: A review. *Bulletin Amer. Meteoro. Soc.*, 80, 1077–1098.

Lowe, P. October 27, 1997. El Niño may have been one factor. *Denver Post*, p. 2.

———. October 30, 1997. Effects of El Niño Cloudy. *Denver Post*, p. A-1.

Lowe, P., and McPhee, M. October 26, 1997. Blizzard of '97: El Niño's first punch, roads closed, thousands stranded. *Denver Post*, p. A-1.

Malmquist, D., and Michaels, A. 1999. Severe storms and the insurance industry. In *Storms*, ed. R. A. Pielke Jr. and R. A. Pielke, London: Routledge Press, in press.

Martinez, J. C. October 24, 1997. Webb vows city ready for El Niño. *Denver Post*, p. A-1.

Martinez, J. C., and J. Hughes. October 29, 1997. Airport director concedes mistakes. *Denver Post*, p. A-1.

Murphy, A. H. 1993. What is a good forecast? An essay on the nature of goodness in weather forecasting. *Weather and Forecasting*, 8, 281–293.

———. 1996. The Finley affair: A signal event in history of forecast verification. *Weather and Forecasting*, 11, 3–20.

———. 1997. Forecast verification. In *Economic Value of Weather and Climate Forecasts*, ed. R. W. Katz and A. H. Murphy. Cambridge University Press. Cambridge, U.K. 19–74.

New York Times. November 9, 1997. Ecological battle over river basins in California, p. 15.

———. February 14, 1999. An analysis of El niño from start to finish. p. 25.

National Weather Service. June1998. *Service Assessment: Central Florida Tornado Outbreak, February 22–23, 1998*. Washington, DC: NWS

National Oceanic and Atmospheric Administration. March 10, 1998. *California Workshop on Regional Climate Variability and Climate Change, Report of Breakout Discussions and Self-Organized Meetings*. Washington, DC: NOAA.

———. June 17,1997. *El Niño plays prominent role in weather patterns for coming seasons*. Press Release 97-37. Washington, DC: NOAA.

———. November 5, 1997. *Strong El Niño likely to impact global weather into spring 1998*. Press Release 97-65A. Washington, DC: NOAA.

———. September 9, 1998. *National Weather Service climate experts make La Niña predictions*. Press Release 98-R235. Washington, DC: NOAA.

———. June 22, 1998. *Spring wraps up warm, with areas of extreme precipitation; La Niña to impact fall and winter weather*. Press Release 98-38. Washington, DC: NOAA.

Pielke, R. A. Jr. 1999. Who decides? Forecasts and responsibilities in the 1997 flooding of the Red River of the North. *Applied Behavioral Science Review*, 3, 34–38.

Pielki, R. A. Jr., and Landsea, C. W. 1999. La Niña, El Niño, and U. S. hurricane damages. *Bulletin Amer. Meteoro. Soc.*, in press.

Pool, B. July 12, 1998. From havoc to hangnails, warm-storm phenomenon took the blame. *Los Angeles Times*, p. B1.

Roebber, P. J., and Bosart, L. F. 1996. The complex relationship between forecast skill and forecast value: A real-world analysis. *J. Weather and Forecasting*, 11, 449–554.

Sarewitz, D., Pielke, R. A. Jr., and Byerly, R. in press. *Prediction: Decision Making and the Future of Nature*. New York: Island Press.

Schrader, A. July 26, 1997. El Niño "blockbuster" brewing, scientists warn unpredictable phenomenon could bring drought or deluge. *Denver Post*, p. B-3.

———. October 28, 1997. It's tough to blame on El Niño, but snowstorm had its footprints. *Denver Post*, p. A-16.

Scheeres, J. August 24, 1997. California braces for predicted fierce winter. *Los Angeles Times*, p. 26A.

Showstack, R. 1997. Scientists aim to translate El Niño data into mitigation efforts. *EOS: Transactions Amer. Geophys. U.*, 78, 477.

Snel, A. November 2, 1997. Webb's DIA woes melt quickly after blizzard. *Denver Post*, p. A-31.

Spotts, P. April 14, 1998. El Niño leaves calling card: History's most costly weather tantrum could cost $40 billion. *Hamilton Spectator*, p. D5.

Stewart, T. 1997. Forecast value: Descriptive decision studies. In *Economic Value of Weather and Climate Forecasts*, ed. R. W. Katz and A. H. Murphy. Cambridge University Press. Cambridge, U.K.

U.S. House Science Committee. September 11, 1997. *Hearing on the state of science used in weather predictions*. Subcommittee on Energy and Environment, House of Representatives, Washington, DC.

Uppenbrink, J. 1997. Seasonal climate prediction. *Science*, 277, 1952.

USA Today. October 22, 1997. California officials prepare for El Niño, p. 2.

Weber, B. October 27, 1997. El Niño not to blame for blizzard. *Rocky Mountain News*, p. 10A.

World Meteorological Organization. June 1998. The 1997/1998 El Niño. *World Climate News*, 13, 5.

8

Summary

*Surprises, Lessons Learned, and
the Legacy of El Niño 1997–1998*

STANLEY A. CHANGNON,
ROGER A. PIELKE JR.,
DAVID CHANGNON, & LEE WILKINS

Much has been said about El Niño 97–98, some of it profound and some not. Several of the key findings from this assessment are reflected in an excellent short summary published by the World Meteorological Organization (WMO) in January 1999.

> The 1997/1998 El Niño was probably the strongest in recorded history; it disrupted the lives of millions of people on all the Earth's inhabited continents. Not all climate extremes and severe weather events of the period, however, can be directly attributed to the El Niño event. Further, not all its impacts were negative, and some regions that were expected to suffer were not seriously affected. (Obasi, 1999)

As the WMO found on a global scale, we have documented the profound impacts of El Niño 97–98 in the United States. But, perhaps contrary to conventional wisdom, the impacts in the United States were, in the aggregate, positive. Because El Niño shifted the geographical distribution of seasonal anomalies and because scientists were able to anticipate these shifts, many decision makers were able to profit from the early warnings to take compensatory actions. The accuracy of the predictions, and the successful use by decision makers of those predictions, offers the promise of the development of a more robust climate service in the United States. The remainder of this chapter summarizes the surprises, the lessons learned, and the legacy of El Niño 97–98.

KEY FINDINGS

Chapter 2: The Atmosphere's Behavior

Once the rapid onset of El Niño conditions was detected late in the spring of 1997, forecasters successfully predicted the event's strength and duration. The oceanic predictions reinforced the ensuing seasonal climate forecasts. The official seasonal outlooks issued by the National Oceanic and Atmospheric Administration (NOAA) in the summer of 1997 skillfully predicted the fall, winter, and early spring 1997–1998 conditions in many parts of the United States many months in advance. The seasonal forecasts had an accuracy of greater than 50 percent for temperatures and of between 30 percent and 50 percent for precipitation, the highest levels of accuracy ever attained, a reflection of the benefits of the considerable research and ocean-monitoring efforts directed at the El Niño Southern Oscillation (ENSO) over the past twenty years. The weather conditions of the fall, winter, and early spring, which included wet and stormy conditions in the South and West and warm and dry conditions in the North, were similar to those experienced in past El Niño events. As one El Niño expert observed, "It is no accident that seasonal predictions made by several research groups around the world for last winter [1997–1998] were quite accurate" (Shukla, 1998).

Chapter 3: The Media

The coverage of the El Niño event by the press and the broadcast media was extensive and sometimes astonishing. In the early months of El Niño, forecasts couched in scientific terms were the primary theme of the news coverage, but this shifted in September–October after a series of weather events, including a damaging October snowstorm that struck Denver and the High Plains, was attributed to El Niño. By November a different news narrative had emerged. No longer were the El Niño predictions scientifically contentious in news reports. Instead, the meta-narrative focused on specific weather events and the problems they created, linked directly to weather patterns either created or exacerbated by El Niño. Thus, heavy rains attributed to El Niño led to mud slides, property damage, coastal erosion, traffic snarls, and school closings along the West Coast. El Niño, in the span of four months, had mutated from scientific predictions worthy of scientific study and debate to climatological fact.

In a popular sense, El Niño arguably became oversold. It became the "cause" of any unexpected event and became a widely known national issue, along with others making headlines, as illustrated in the cartoon in Figure 8-1. If popular culture is taken as one measure of human concern with a particular issue, then El Niño clearly captured the public's imagination. The original news story image of El Niño had been transformed into something more symbolically charged. Contentious scientific predictions and uncertainties became believable warnings, which, in turn, were transformed into a single cause responsible for multiple effects. El Niño was fact, and it was accepted without question as the cause of an enormous variety of weather-related consequences. It sent a signal throughout society that something important—and far-reaching—was changing. A new concept of weather and climate had emerged. The news coverage of El Niño may symbolize much more than what happened between May 1997 and June 1998, but the value of the issue's salience may have been diminished by its over-exposure in the media.

Chapter 4: The Scientific Issues

Three issues predominated in U.S. science concerning El Niño 97–98. Of greatest scientific importance are the official seasonal climate predictions issued during the summer of 1997. They defined most of the critical anomalies in the United States that were experienced in late fall, winter, and early spring 1998. These were the most accurate long-range predictions ever made for the United States. Most experimental forecasts of this period agreed with the official forecasts.

The second important scientific issue raised by El Niño 97–98 involved the physical and socioeconomic impacts expected as a result of the forecast seasonal climate conditions. These "impact predictions," which were made primarily during the late summer and fall of 1997, were inconsistent and speculative and differed in major ways. They were issued by a wide variety of individuals in the public and private sectors. Their marked differences (e.g., about the economic effects of a mild fall season) led to confusion in the minds of many decision makers and among the public over how and when to react to the climate predictions. Doomsday outcomes became an overriding theme, often promoted by govern-

FIGURE 8-1.   El Niño 97–98 became a household word for many months and constantly figured in headlines in the national news. This cartoon, which was widely published in early March, reveals that the El Niño event ranked alongside another story of major national interest. (Reprinted with permission of Don Wright)

ment agencies, many scientists, and the media. These wide-ranging differences over expected weather impacts reflect the fact that most people and professionals do not understand well the complex effects of weather on the nation's social, technological, and environmental systems.

The third major scientific issue to emerge during the El Niño 97–98 event involved the event's possible relationship to global warming. Some scientists claimed that the strong El Niño resulted from the effects of an enhanced greenhouse effect on the atmosphere, whereas others openly challenged this claim, arguing that there was no relationship between the two. Caught up in this issue of a possible relationship between global warming and El Niño are scientific claims that the resulting cold-season weather conditions of 1997–1998 resembled those to be expected under future global warming, a "window on global warming." In the years prior to 1997, the global warming issue had been fraught with scientific and political implications, and the scientific controversy over the relationship of El Niño and global warming fit right into the long-standing global warming debate.

Chapter 5: Uses of the Predictions

The assessment of the uses of the El Niño–based seasonal climate forecasts by eighty-seven individuals, all decision makers in weather-sensitive institutions (agribusiness, water management agencies, power utilities, insurance, transportation agencies, and emergency management), revealed important information. Nearly 50 percent used the forecasts in making long-term planning or operations decisions, a much higher level of usage than found in prior assessments of long-range forecast utilization. Furthermore, at the time of the interviews, 93 percent of the users reported beneficial outcomes from the applications of the forecasts. The applications of the forecasts were quite diverse. The corporate and consumer gains reported by the power utility sector varied from $200,000 to $30 million per utility.

The user survey found differences among economic sectors in the use of forecasts, in their applications, and in the problems that restricted their use. About 60 percent of the water managers and utility decision makers we interviewed used the climate forecasts, with most applications in water resources related to planning and in utilities to operations. Many water managers were risk averse and decided not to use uncertain seasonal forecasts in their operational decisions, whereas deregulation of the utility industry has increased competition, leading to a greater operational use of forecast information. The agribusiness and crop insurance officials we sampled found little use for the cold-season forecasts, since their primary concern is growing-season conditions. The major impediments to usage of the forecasts among all sectors included concern over the accuracy level, lack of desired information such as in-season extremes (e.g., number of storms), and the differences found between the official forecast conditions and those being issued by other institutions. Part of this problem relates to the Internet and the wide availability of information. Anyone can offer predictions and disseminate them widely. The Internet allows anyone to be an "expert" on any topic, including weather forecasting.

Chapter 6: Impacts of the Weather

It was "the winter of El Niño's discontent" (Schmid, 1998). The predictions of cold-season climate conditions for 1997–1998 and the impacts expected as a result of El Niño 97–98 weather conditions focused almost totally on bad weather and negative outcomes. Indeed, the wet and stormy winter weather in California and in the Deep South created major damages, leading to estimated losses of about $4 billion. California was assaulted by coastal storms and heavy rains, which caused floods and numerous landslides, as well as damage to the state's valuable agricultural output, with losses put at $1.1 billion statewide. Florida, Texas, Georgia, and several other southern states were struck by severe rainstorms and numerous tornadoes. Storms led to more than one hundred deaths, and El Nino–related property and agricultural losses in Florida ultimately reached $500 million. A record early-season damaging snowstorm swept across the High Plains and the upper Midwest in October, and then an extremely severe ice storm struck the Northeast in January, creating losses in excess of $400 million and twenty-eight deaths. By the end of May 1998, the national death toll caused by weather related to El Niño was 189. The nation's major economic losses included property losses of $2.8 billion, government relief totaling $460 million, and agricultural losses of $700 million. Not all scientists agreed that the damaging storms were due to El Niño, but our accounting of losses includes all those events that some scientists, speaking in an official capacity, attributed to El Niño's influence on the weather conditions.

In contrast, the mild, almost snow-free winter in the northern United States resulted in major gains and benefits. Many fewer deaths occurred due to winter conditions (e.g., bad roads, low temperatures) than normally occur. Deaths caused by poor winter conditions fell dramatically, with 830 fewer deaths occurring than in an average winter. El Niño's influence contributed to the elimination of major Atlantic hurricanes during 1997, an outcome that produced enormous savings to home and business owners, the government, and insurers; average annual hurricane losses are $5 billion in property damages and twenty deaths. The warm winter led to major reductions in heating costs because of reduced use of natural gas and heating oil, for a savings of $6.7 billion to consumers.

Many people changed their normal winter behavioral patterns, and thousands went outdoors more. Millions went shopping, many people altered their types of recreation, and almost everyone enjoyed better health than in normal winters. There were notably fewer airline and highway transportation delays due to inclement weather and airline profits increased 15 percent because of the better winter weather and fewer costly delays. The lack of snowfall and freezing rain led to major reductions in the use of salt on highways and streets and reduced labor costs, creating major savings to state and local governments. The generally good weather also had a major influence on construction, retail shopping, and home sales at a time when the nation's economy was robust. Many retail chains reported record high sales for January through March, and record-high sales of homes occurred over a four-month period. The federal government had lower relief costs than in prior winters. In addition, many benefits came to those who used the El Nino–based long-range predictions for winter and spring weather conditions. Benefits were derived from the widespread mitigation activities in California, for example. The difference in losses in California for two similar El Niño events is sizable: $2.2 billion in 1982–1983 (adjusted to 1998 dollars) versus $1.1 billion in 1997–1998, suggesting the extensive mitigation activities, which cost an estimated $165 million, were effective.

The net benefit to society and to the nation's economy from these varied gains amounted to nearly $19 billion. The headlines in Figure 8-2 describe several of the major impacts, both good and bad, of the El Niño weather conditions. The fact that El Niño led to net benefits, rather than costs, will surprise some. The United States, compared to many other nations affected by El Niño 97–98, is an overall climate "winner." Yet, there remain numerous approaches to reducing still further El Niño losses and to taking advantage of the forecast opportunities when future events occur.

Chapter 7: Policy Responses to Predictions

Scientists did an impressive job of predicting a range of climate anomalies in the United States, ranging from excessive rainfall and storms in California and Florida to depressed hurricane activity in the Atlantic. The endeavor represents a step toward "operational forecasting" for seasonal climate conditions, and toward the development of a robust program of climate services based on use of climate predictions. However, for society to realize the benefits related to climate predictions, attention will have to be paid to both the production of forecasts and the use of forecasts by decision makers. It is also important to realize that a comprehensive evaluation of skill and value of any forecasting process requires a considerable research effort. Four cases of applications of the El Niño 97–98 predictions revealed major lessons relating to the forecast process, involving the generation of the prediction, the communication of the forecast, and the use of the forecast.

The use of climate predictions to reduce losses in California seems like a best-case scenario, but important lessons were illustrated. First, the effective use of a skillful forecast does not always mean benefits—in one instance, the predictive information led to costs, rather than benefits, from the perspective of those want-

Meteorologist sees weather woes ahead

El Nino gets the blame for weirdness

Storms

Clinton visits twister-torn Florida; 3 still missing

But the weather 'event,' while strong, is only part of the reason for a rash of unusual weather that has befallen the country so far

Mud brings disaster in California

El Niño 1997-98

Changing the Way We Think About the Weather

Energy prices plunging

Consumers, economy benefit as oil supply surpasses demand

Winter damage from El Nino bad, not worst

Storms Are Evidence—but Not Proof—of El Niño Connection
■ Weather: Climate experts say they can't analyze effects of the warm-water pattern until they can study the whole season.

FIGURE 8-2. Headlines illustrating some of major outcomes and certain reactions resulting from the El Niño–related weather conditions during 1997–1998. These reflect both the positive and negative outcomes experienced.

ing to preserve the flora and fauna in the Los Angeles floodways. Often, an assessment of the value of predictive information must include consideration of who benefited and who did not, rather than just aggregate outcomes. Second, forecasters must be careful to manage the expectations of decision makers. The California case was a huge success, and forecasters have not been shy in touting their success. But, what would have happened if the 1997–1998 storms had failed to materialize? All predictions of climate are necessarily probabilistic, meaning there will be times when a skillful prediction is made, yet the predicted event does not occur. Hence, an important challenge for forecasters is to manage expectations following realized forecasts.

In the Colorado case, other lessons were taught. A seasonal prediction for heavy spring snowfall had value in encouraging regional officials to prepare for an early October blizzard because there was room for improvement in Denver's snow removal plans. The climate predictions stimulated a local "no regrets" response that would have made sense without the predictions but that probably would not have occurred without the predictions. Even a skillful prediction of the blizzard would not have addressed the problems involving access to the city's airport. The lesson is that the strengths and weaknesses of existing decision processes are a key factor in whether a forecast leads to societal benefits.

The case of the tornado outbreak in Florida presents another important lesson—that skillful predictions alone are an insufficient condition for society to realize benefits. An effective forecast process involving the user is necessary if predictive information is to be well used and if benefits are to result.

The Federal Emergency Management Agency's (FEMA) apparent misunderstanding, its equating El Niño with greater disaster costs nationwide, suggests that the scientific community was at once too successful and not successful enough in publicizing the coming El Niño. It was too successful to the extent that people came to associate El Niño with Armageddon, and not successful enough to the extent that the public failed to appreciate the subtleties of extreme weather related to the event. If the media and the public fail to accurately understand El Niño, much less the El Niño–based climate predictions, then chances are decreased that even skillful forecasts will have value to decision makers. These are all very important lessons to learn as the nation develops skill in seasonal climate forecasting.

WINNERS AND LOSERS
Society

El Niño 97–98 led to mild winter weather over much of the nation, creating better health and happiness, many more lives saved than lost, and a low incidence of respiratory diseases. Millions reported enjoyment of the mild winter weather and took advantage of the opportunity to be out of doors often. The low heating costs saved consumers billions of dollars, and the mild winter weather, few winter storms, and lack of Atlantic hurricanes saved an estimated 850 lives. Conversely, numerous tornadoes and floods led to 189 deaths.

Economy

Weather effects on transportation, retailers, consumers, insurers, the government, construction industry, and snow removal groups produced major winners. Losers included agriculture and property owners in some areas and snow-dependent industries in parts of the nation. Effects on the tourist and recreation industries were mixed. Major winners also included institutions that chose to use the fall, winter, and spring climate forecasts. The survey of decision makers in weather-sensitive positions revealed that most forecast applications led to beneficial social, economic, and environmental outcomes. In some instances, the financial gains were sizable. The major losses occurred in California, Florida, and other southeastern states.

Atmospheric and Oceanographic Sciences

The survey of eighty-seven users of the seasonal forecasts based on El Niño conditions revealed that the forecast institutions, including CPC, were given high marks by most and that their credibility was much improved among knowledgeable users. The sampling further revealed that there was strong evidence that the general public also became better informed and more knowledgeable about weather and climate and about the influence of the oceanic conditions on weather. Furthermore, the public became more interested in what the future climate will be and in the climate change issue.

National Oceanic and Atmospheric Administration

The survey of users of the Climate Prediction Center (CPC) predictions revealed that many (64%) felt that the forecasts had greatly enhanced the reputation of NOAA and CPC. The positive reactions to NOAA, based on its forecasts and on the wealth of useful El Niño information it made available, were further revealed in the analysis of hundreds of news stories. There is a major opportunity for NOAA to exploit this "victory" with follow-up activities that respond to this new attention to long-range forecasts. But, at the same time, this situation presents a challenge for NOAA forecasters to meet the expectations of decision makers. Not all future forecasts will have the success of those based on El Niño 97–98. A critical challenge for NOAA will be to effectively communicate probabilistic information.

MAJOR SURPRISES

El Niño 97–98 yielded a number of unexpected outcomes. These include the following:

- The rapid development of a strong, record-breaking El Niño went unpredicted.
- The 1997–1998 event became the largest El Niño ever recorded.
- Winners exceeded losers, in terms of human lives and dollars saved, in the United States.
- Highly accurate seasonal climate forecasts were based on El Niño.
- Major positive reactions from users of the seasonal climate predictions were noted.
- The media and the public were smitten by El Niño, and it became a household word.
- Mitigative actions produced large and measurable savings.
- The winter conditions mimicked those expected from global warming.

LESSONS LEARNED

Lesson 1   NOAA needs to more effectively coordinate its release of information about El Niño and La Niña-type weather conditions in order to minimize differences expressed by various staff. NOAA also should assess who speaks during such events and about what topics. It needs to keep its field staff informed about the effects of ENSO events on severe storms (chapter 4).

Lesson 2   There is an opportunity for NOAA to exploit its forecasting success and its new visibility and credibility. NOAA should emphasize the "official" forecasts over others issued by scientists and institutions issuing "experimental" predictions without the appropriate caveats. Outreach endeavors, such as the California pilot project and annual regional workshops, will help educate users and help explain reliable sources of forecasts and why various sources often predict different outcomes (chapters 5 and 7).

Lesson 3   Seasonal forecasts will be much more useful if (1) they contain more information on existing and past forecast accuracy and compare current and past outcomes (e.g., comparing an existing El Niño event with what happened during the last comparable event, such as the 1982–1983 event); (2) they present current probabilistic information about forecast conditions; and (3) they present "climate profiles" of weather conditions during a predicted warm, cold, wet, or dry season (e.g., how many days in a forecast season will have temperatures exceeding various thresholds or the number of storms producing two-inch rainfalls) (chapters 2, 5, and 7).

Lesson 4   El Niño 97–98 created an important opportunity for private-sector meteorological firms. Our sampling of eighty-seven decision makers revealed that meeting their needs often requires having a working relationship with the forecaster to understand the complexities of the weather-related decisions, to obtain tailor-made forecast products, and to have access to expertise (chapter 5).

Lesson 5   Sampling of users of forecasts in emergency preparedness agencies in California, Florida, and other states revealed that mitigation efforts prior to the winter extremes paid off, with reduced flooding and reduced environmental losses (chapters 5 and 7). The losses experienced in California were much less than those during the comparable 1982–1983 El Niño, suggesting that the mitigation efforts worked (chapter 6). A careful study of the California case would provide the government with definitive information on the value of mitigative activities.

Lesson 6   Criticisms were leveled by sources in news stories at scientific forecasters who focused largely on negative outcomes expected from the El Niño weather, particularly when it became apparent that the aggregate gains from the national weather conditions would be quite large and exceed the losses. Dire warnings from weather experts and government agencies can be misleading and hurt credibility when different outcomes occur. It is important for forecasters to predict winners as well as losers (chapter 6).

Lesson 7   After the El Niño–based seasonal climate forecasts were issued during the summer of 1997, a plethora of impact predictions were issued by a number of scientists and persons in various businesses. Many of the predicted outcomes disagreed, many were incorrect, and many oversimplified weather effects. This revealed a major lack of understanding of how weather conditions impact society, the economy, and the environment (chapter 4).

Lesson 8   Accurate seasonal climate predictions, even though they remain scientifically problematic, will be recounted as fact once an event is under way and actual weather outcomes fulfill the predicted events (chapter 3).

Lesson 9   Agencies in the business of issuing forecasts and warnings may have to deal with unintended consequences of success, including inflated expectations and the resulting fallout when, inevitably, a long-range prediction about future weather turns out be just that: a prediction. Furthermore, subsequent predictions, for example, those based on the 1998–1999 La Niña phenomenon, may receive less critical journalistic examination and policy attention because of the perceived accuracy of the 1997–1998 El Niño forecasts (chapter 3).

Lesson 10   When weather conditions or climate events become "national news," news and science writers tend to approach "local" experts for interpretations of the events and conditions. These experts often provide widely different explanations on questions such as what tornadoes to attribute to El Niño or whether global warming will occur or has begun (chapters 3 and 4).

Lesson 11   Climate research, including modeling, observations, and historical understandings, was an important factor in the success of the predictions made in El Niño 97–98. But, the issues associated with predicting the onset and duration of the event revealed shortfalls in scientists' current understandings. In a related vein, debate and discussion of the possible link between the intensity and frequency of El Nino events and global climate change revealed uncertainties in our understanding of the effects of longer-term climate variability. Additional research will help to address these scientific issues and contribute to improvements in the climate forecasting process (chapter 4).

Lesson 12   Since the unusual cold season conditions of El Niño 97–98 may be indicative of conditions under future climate change, an assessment of the impacts and reactions to El Niño 97–98, and the effectiveness of the various adaptive actions, would provide information useful in helping scientists understand how society can and should adapt to climate change (chapters 3 and 6).

LEGACY

The most intriguing question about El Niño 97–98 concerns its legacy. The enormous news coverage and the great public interest in the event raise some important questions for both scientists and policy makers. Lingering questions include:

1. What is the relationship between El Niño and global warming?
2. Will the 1997–1998 event be regarded as a signal event creating new belief, in a cultural and political context, in the reality of climate change?
3. Will the increased credibility of climate forecasts lead to rising expectations for those who make predictions, and can forecasters fulfill this promise?

Primary among these is the long-term impact of the accurate climate predictions based on El Niño. The media can aid in improving public understanding of symbols of risk that do not always match scientific findings but that may have important social and political outcomes. While on one level the scientific community should welcome such wide public acceptance of a prediction, the fact that the El Niño–based predictions, at least as they were transmitted in the mass media, were so accurate raises other important issues—will subsequent predictions receive less scrutiny from the popular press and policy elites than is warranted? Will subsequent, less accurate predictions create disillusionment among users and the public?

One of the potential legacies of El Niño 97–98 concerns a major future change in the use of long-range climate predictions. The accuracy of the predictions during the event and the major economic gains resulting from the use of the forecasts present an opportunity to improve the wise use of forecasts in the future. This will involve improving the three processes in the successful use of predictions: the prediction process of the provider, the communication process between the provider and user, and the choice process of the user. Thus, improved accuracy, better communication, and greater user understanding are keys to the process. The roles of the public-sector prediction providers, including the numerous scientific research entities generating "experimental" forecasts, and the private-sector providers in this three-phase process need to be better defined to enhance wise usage of predictions. El Niño 97–98 has created a new "market," and new uses of long-range forecasts will evolve in time and across application areas at a rate that depends on how well the processes are handled.

Another potential legacy of El Niño 97–98 relates to the impression it has left among the American public and policy makers. If the public begins to believe that such events as El Niño are harbingers of future global climate change, will that belief fuel other sorts of preparedness and adaptive activity? The totality of the news coverage suggests that mediated risk communication about El Niño could have functioned as a signal event regarding the larger issue of global climate change. Such a signal could influence both public understanding and political activity related to climate. Will El Niño become the signal event for global warming, as chapter 3 suggests? Or, will this connection be another example of El Niño "hype," as suggested in Chapters 4 and 7? Time will tell.

REFERENCES

Obasi, G. O. P. 1999. El Niño in Review. *World Climate News*, 14, 3.

Schmid, R. E. May 7, 1998. El Niño not to blame for weird weather—victims of every climate pattern invoke the name of the seasonal anomaly. *Detroit News*, p. A15.

Shukla, J. 1998. El Niño and climate more predictable than previously thought. *Bulletin Amer. Meteoro. Soc.*, 79, 2816.

INDEX